Sensational
Vegetable
RECIPES

The Confident Cooking Promise of Success

Welcome to the world of Confident Cooking,
where recipes are double-tested by our team
of home economists to achieve a high standard
of success—and delicious results every time.

bay books

C O N T E

Sweet Spiced Golden Nuggets, page 80.

Warm Tomato and Herb Salad, Page 108.

Vegetable Basics	4	Side Dishes	92
Snacks	12	Index	111
Quick One-Step Recipes	28	Glossary	112
Soups & Starters	48	Useful Information	112
Main Courses	68		

Red Capsicum Soup, page 66.

Vegetable Samosas, page 26.

N   T   S

Gourmet Vegetable Pizza, page 78.

Honeyed Baby Turnips with Lemon Thyme, page 109.

The Publisher thanks the
following for their assistance
in the photography for this
book.
Barbara's Storehouse; Hale
Imports; The Pacific East
India Company; Perfect
Ceramics; Villeroy & Boch;
Waterford Wedgewood

The Publisher thanks the following for their assistance in the photography for this book. Barbara's Storehouse; Hale Imports; The Pacific East India Company; Perfect Ceramics; Villeroy & Boch; Waterford Wedgewood

The test kitchen where our recipes
are double tested by our team of
home economists to achieve a high
standard of success and delicious
results every time.

When we test our recipes, we rate
them for ease of preparation. The
following cookery ratings are on
the recipes in this book, making
them easy to use and understand.

A single Cooking with Confidence
symbol indicates a recipe that is
simple and generally quick to make
– perfect for beginners.

Two symbols indicate the need
for just a little more care and a
little more time.

Three symbols indicate special
dishes that need more investment
in time, care and patience—but the
results are worth it.

IMPORTANT
Those who might be at risk from
the effects of salmonella food
poisoning (the elderly, pregnant
women, young children and
those suffering from immune
deficiency diseases) should
consult their doctor with any
concerns about eating raw eggs.

Two-cheese Risotto Cakes, page 57.

Spinach and Salmon Terrine, page 53.

Vegetable Basics

Our markets and supermarkets are overflowing with delicious and healthy vegetables. This guide to purchasing, storing, cooking and presentation will help you prepare successful and delicious vegetable dishes.

Vegetables picked fresh from the garden are still the best but, as most of us do not have a vegetable garden to harvest, we turn to the local greengrocer or supermarket. And there, thanks to modern methods of cold storage and transportation, we find vegetables available to us year-round that a couple of years ago were only on sale for a few short weeks. The sheer abundance and variety of vegetables, whether fresh, frozen or packaged, means that the days of meals featuring boring, boiled vegetables are banished forever.

This book brings together all the basic methods of storing, preparing and cooking vegetables to retain their colour, flavour and nutrients, plus recipes from many nations. For the adventurous cook, Asian food stores and vegetable markets are full of delicious ingredients, all quite simple to prepare.

Purchasing

Little and often is a good maxim to follow when shopping for vegetables. Always buy vegetables that look fresh and crisp and have a bright natural colour. Buying fresh vegetables in season is economical, and means you get the best flavour because of the short storage time.

If you have a definite vegetable recipe in mind, but can only find sad-looking specimens of the one you require, find a substitute (e.g. canned tomatoes or frozen spinach) or buy another vegetable that looks in peak condition, and save your first choice until you can find a fresh supply.

Freshness is important, especially in green leafy vegetables with a high water content, such as lettuces and spinach, because vitamin losses begin soon after picking.

Many vegetables are available ready-packed in net or plastic bags, or sometimes polystyrene trays, in convenient amounts. Before buying these, check that the produce is fresh and undamaged.

Roots and tubers—carrots, potatoes, parsnip, beetroots, turnips—should be unblemished, with no musty smell. Loose, unwashed potatoes are a better buy than washed and packaged ones; you can pick them over for size and quality, while the earth around the potato keeps it in good condition. Don't buy any that are sprouting or have a green tinge.

Onions, white, red or brown, are at their 'sweetest' when their tops are still green, but you won't find them like that in the shops, as they are picked when their tops have shrivelled to allow maximum growth. Choose those with smooth skins and no dark or damp patches.

The Brassica family (all types of cabbage, Brussels sprouts, broccoli and cauliflower) should have crisp leaves; cabbages should be tightly packed and heavy; cauliflowers should be white or creamy-white, without any discoloured patches. Broccoli should have firm, tight, dark green florets, not yellowed or loose-looking; avoid any that have flowered.

Buy eggplant (aubergine) that are firm, with shiny smooth purple skin; capsicum (pepper) (particularly check the red ones) and cucumbers should be crisp and quite hard, with no wrinkled skin or mushy patches—if in doubt, squeeze gently.

Tomatoes can be purchased ripe or not, according to your preference and how they are to be used. Whether you are buying large meaty types or tiny cherry tomatoes choose the reddest you can find, as firm or soft as you need

To store mushrooms, wipe over with paper towels or a damp cloth.

Place into a brown paper bag and refrigerate for up to three days.

To store lettuce, wash, dry and loosen core by hitting on a bench or board.

Turn the lettuce over and twist out the core. Wrap and refrigerate.

Storing

Shopping frequently for vegetables means that you can take advantage of what looks best on the day, but unfortunately this is not always practical. Shopping weekly means that you must store your vegetables carefully to retain maximum flavour and vitamins.

Most vegetables benefit from being stored in a cool, dark place. Generally, vegetables keep best when stored in the crisper section of the refrigerator. Store them unwashed and loosely packed in plastic bags. Squeeze the air out of the bags before storing. Fresh ginger can be refrigerated unwrapped.

(vine ripened have the best flavour). Leave firm tomatoes to ripen at room temperature. Keep in mind that canned tomatoes are perfect for many dishes.

Legumes (green beans, broad/fava beans, etc, and peas, sugar snap peas and snow peas/mangetouts) should be bright green and have no wrinkles in the skin. Smaller specimens are younger and therefore more tender. They should 'snap' rather than bend when broken.

Sweet corn is also best when young. This summer treat should be delicious enough to eat raw, but usually has become a little too tough by the time it reaches the shelf. Choose cobs with unblemished husks. Pull the husks back to check that the kernels are even-sized, small and tender, and that there are no caterpillars lurking. Dented-looking kernels mean the corn is not fresh and will be tough when cooked.

Freshness is particularly important with salad vegetables. Don't buy green vegetables that are wilted, dry-looking or have yellow or brown patches or insect-nibbled leaves. Crisp-head lettuces (iceberg, cos/romaine) should feel firm when squeezed and the base should be dry.

Many loose-head lettuces are sold with their roots intact; this helps to keep them in good condition.

Stalky vegetables, such as celery, fennel and asparagus, should be crisp and firm with no brown patches. Celery should be heavy and stand up straight, with light green leaves. Fennel should be white and crunchy.

When buying asparagus, choose straight even-sized stalks with tight buds—thick stalks are more tender than thin. The cut end should be dry but not withered.

Artichokes should have silky, compact green heads with no dark patches.

Choose avocados according to what you are using them for. For example, you may want a really ripe one to purée, a perfect one to eat immediately or a firmer one to keep. Press gently to test for ripeness; the flesh should just give at the stem end. Avoid those with bruised or black/brown patches. Place hard avocados in a brown paper bag and put in a warm place (e.g. on top of the refrigerator, at the back over the motor) to ripen in two days.

Bunches of fresh green herbs such as parsley, chives and basil are best kept in water in the shop, and should look and smell freshly picked.

Leafy green vegetables such as lettuce and spinach should be washed and thoroughly dried by spinning or patting dry with a tea towel or paper towels. (Water leaches out vitamins from the leaves.) Hit the base of the lettuce hard on a bench or board to loosen the core, then turn over and twist it out. Pack whole lettuce or leaves loosely in plastic bags. Squeeze air out of the bag, seal and place in the crisper section of the refrigerator. Properly stored, lettuces will last up to seven days for crisphead and two to three days for soft butter types.

Mushrooms can be stored in the refrigerator, but should be wiped clean and placed in a brown paper bag. This way they will keep in good condition for three to four days. Do not store them in plastic because they will begin to decay quickly.

Potatoes should not be refrigerated as they can develop a sweetish taste. To store, remove from plastic bags and place in a hessian or paper bag. Keep them in a cool, dark, dry place with good ventilation. Properly stored, they should remain in good condition for two months or more. Sweet potatoes, turnips, onions (except for spring onions/scallions) and garlic should be stored in the same manner.

For convenience, both raw garlic and ginger can be puréed in a food processor or blender and stored in glass jars in the refrigerator.

Tomatoes are a subtropical crop and low temperatures damage their cell structure, so they should not be refrigerated. The optimum storage temperature for tomatoes is 10°C, but keeping them at room temperature until needed is the most convenient for

To prepare asparagus, break off the hard ends and discard.

Trim any remaining woody ends with a vegetable peeler.

most cooks, and underripe fruit will ripen best at room temperature. Take tomatoes out of plastic bags to avoid the growth of bacteria. Keep them away from sunlight—it destroys vitamin C. Store in a single layer unless you want the tomatoes to ripen quickly, as heat rather than light causes ripening. Remove any tomatoes that have spoilt.

Freezing

Most fresh vegetables are not suitable to freeze raw, but many can be blanched, then successfully frozen. Blanched asparagus spears, beans, broccoli, carrots, cauliflower, corn (whole cobs or kernels), shelled peas and skinned tomatoes all freeze well. Freezing is a good way to take advantage of a seasonal glut in produce.

Pick over and discard any damaged specimens. Wash and cut the vegetable into pieces as desired. Plunge into boiling water for 30 seconds, then into iced water to refresh. Pat dry with paper towels, pack into freezer bags, expel air and seal. Label and freeze. Frozen blanched vegetables will keep for up to three months.

Fresh herbs can be successfully frozen. Wash and dry the leaves, place into small freezer bags, expel air and seal. Label and freeze.

To freeze cooked vegetable dishes, stock, purées, soups and sauces, cool the just-cooked dish quickly, place in plastic containers, seal, label and freeze. As a general rule, freeze for up to two months.

Commercial or home-frozen vegetables should be added to recipes unthawed, and toward the end of the cooking time.

Stocks, sauces and precooked dishes should be thawed in the refrigerator and reheated gently before serving.

To salt eggplant (aubergine), place it in a colander, sprinkle over salt.

Preparing

Before cooking, wash vegetables well to remove any pesticide residue, as well as grit or insects. Leeks, spinach and silverbeet (Swiss chard) need to be washed thoroughly to remove sand. Some vegetables require particular techniques to cook and present them at their best. The following methods are frequently used in this book.

Salting. Salting is used when cooking eggplant (aubergine) and large zucchini (courgette) or cucumbers, to draw out excess moisture and any bitterness from the vegetable. Salting eggplant slices before frying will cut the amount of oil needed by two-thirds. To salt vegetables, chop or slice according to the recipe. Place in a large colander or spread on paper towels and sprinkle with salt. Leave for at least 30 minutes, rinse under cold water, drain and pat dry with paper towels.

To peel tomatoes, make a small cross on the bottom of each tomato, and plunge into boiling water for 30 seconds. Lift out and plunge into chilled water for one minute, using plenty of water in each case. The skin can then be peeled easily from the cross using a downward motion.

After 30 minutes, rinse under cold water and pat dry with paper towels.

To prepare artichokes, trim the stalk from base of artichoke. Using scissors, trim hard points from the outer leaves. Using a sharp knife, cut top from artichoke. Brush all cut areas with lemon juice to prevent discolouration. Cook artichokes in a stainless steel or enamel pan.

To prepare asparagus, break off the inedible woody ends and discard. If the ends are still hard, trim with a vegetable peeler.

To prepare avocado, cut the avocado in half using a sharp knife. To remove the stone, push the knife into the stone and gently twist and lift out. The skin can be peeled off with fingers.

To prepare cucumber. When seeding is called for in the recipe, peel the cucumber using a vegetable peeler (if peeling is specified), scrape out the seeds with a teaspoon and discard. Cucumbers can be given a decorative finish by peeling in stripes, or by running the tines of a fork down the outside after peeling. Use as directed.

To crush garlic cloves. Peel the garlic and place on a cutting board. Using a very sharp knife, chop the garlic finely, working in a little salt as you go. Scrape the chopped garlic together into a mound. Turn it over

To peel tomatoes, make a cross cut in each tomato. Plunge into hot water.

Plunge into cold water, then peel skin down from the cut.

To chop onions, make five or six cuts through, turn and slice across.

To prepare artichokes, cut off ends with a sharp knife.

Trim leaves with scissors and brush cut ends with lemon juice.

To remove avocado stones, place knife blade firmly into stone.

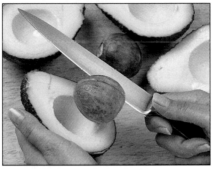

Gently twist the knife and lift out the stone.

with the blade, chop again. Repeat until the garlic is a fine mass. For coarsely crushed garlic, press down heavily with the flat side of a wide cook's knife or cleaver. Cutting the gar-lic clove lengthways makes it easier to remove the skin.

To prepare mushrooms. Never peel or wash mushrooms. After wiping clean, remove stems if they are woody (reserve to make stock) and slice. Peeling and washing mushrooms removes flavour and makes them waterlogged.

To prepare onions. Peel and place on a chopping board. To chop, make five or six slices through the onion. Turn it 90°, hold layers together firmly and cut across them.

Cooking methods

Vegetables can be cooked in many ways, ranging from burying them in the ashes of a campfire, to combining with eggs and liqueur to make a most sophisticated soufflé.

The three cooking methods that have a particular significance for vegetables are blanching, steaming and stir-frying. These are all quick-cooking and retain all the colour, flavour and nutrition of the vegetables.

Blanching. This method is used to precook (or parboil) vegetables before adding to other cooked dishes or salads, or before freezing. The vegetables are plunged briefly into boiling water and then refreshed in cold water. This method preserves the green colour.

Steaming. Vegetables are cut into even-sized pieces and cooked in a basket or on a rack over a little boiling water or stock in a tightly covered pan. A few minutes steaming (depending on size) is enough to make green vegetables tender while remaining firm and flavourful (root vegetables take longer). This method is preferable to boiling because the flavour is preserved as the vegetable juices are not lost into the cooking water.

Stir-frying. Heat a little oil in a large frying pan or wok. Have all the vegetables cut or divided into thin even-sized pieces. Cutting celery, carrots, etc, on the diagonal gives a larger surface area and enables quicker cooking than straight sliced vegetables. When the oil is hot, add the vegetables. Toss and stir briskly around the pan for a few minutes, using a large spatula (not plastic) or spoon, over high heat. Do not let the vegetables sit still in the pan, or they will burn.

Never soak vegetables in water prior to cooking or add soda to the cooking water, as both these methods destroy vitamin C.

Vegetable Juices

A quick and healthy way to make the most of vegetables is to extract the juice from them. Vegetable drinks, alone or combined with other vegetable and fruit juices make an ideal vitamin-packed pick-me-up, instant breakfast, or can be used to add extra vitamins and minerals to your diet.

Wash or peel vegetables and cut into chunks that will fit into your juice extractor. To minimize vitamin loss, cut vegetables immediately before juicing and drink the juice as soon as it is made.

Vegetables suitable for juicing are carrots, celery, tomatoes, parsley, cucumber, capsicum (pepper), cabbage, beetroot and onion. Carrot juice has a smooth and velvety texture and soothes a troubled stomach; cabbage and onion juices are reputed to help those suffering from colds; parsley, cucumber and celery juice purify the blood, while cucumber is also said to help remove cellulite.

Try these delicious juice combinations: carrot and apple; carrot, celery and apple; tomato and celery; tomato, onion and parsley; carrot, beetroot and celery.

Canned tomato juice makes the basis for an instant gazpacho: place chopped capsicum (pepper), chopped tomato, chopped cucumber, chopped celery, a little chopped onion, a crushed garlic clove and chopped parsley in food processor or blender, process until almost smooth. Stir in chilled canned tomato juice to the correct consistency. Season to taste with salt, pepper and a little lemon juice and serve.

ARTICHOKE: Although there are two types of artichoke—globe (pictured fresh and canned), and Jerusalem, the globe is the most often seen. Simply trim off outer leaves, steam for 30 minutes, then peel off the leaves one by one and dip them in melted butter, vinaigrette or hollandaise. The Jerusalem artichoke is a root vegetable from the sunflower family, which can be included in stews and soups, puréed, steamed or fried, or eaten raw in salads.

ASPARAGUS: Held in high esteem in Europe, the white type is rarely seen elsewhere. Green asparagus is available in late spring and in summer. Asparagus is a very light, delicate vegetable and thus lends itself well to rich sauces, such as hollandaise. It can also be served cold in salads, or on its own with a simple vinaigrette or mustard dressing.

AVOCADOS: The avocado was first introduced in restaurants, crowned with fresh prawns and cocktail sauce, or served simply with vinaigrette dressing. It has long since proved its versatility, chopped and tossed in salads, sliced on sandwiches, mashed with garlic and lemon to make guacamole dip. It is actually a fruit, and in parts of South-east Asia is eaten with sugar as a dessert.

BASIL: This herb has a uniquely strong flavour and scent. It is traditionally and extensively used in Italian cooking in tomato-based sauces and as the main ingredient in pesto. Add whole or chopped leaves to salads, soups and pasta dishes.

BEANS: Despite the wide variety in the bean family—French, runner, purple (which turn green when cooked), yellow wax, snake—the green bean (pictured) is most popular. Beans may be topped and tailed and strings removed, sliced thinly with a bean slicer or cut diagonally into short lengths. Use in casseroles and stir-fries, or blanch for use in salads.

BEETROOT: A hardy root vegetable, beetroot is celebrated in the Russian soup, borsch. It also lends itself to pickling in vinegar, spices and brown sugar and is used in salads. Baby beets can be served whole as a vegetable topped with sour cream. Raw beetroot has a pleasant nutty taste; the leaves can be cooked in the same way as spinach.

BROCCOLI: A cousin of the cauliflower, broccoli can be white, purple or green, the latter most commonly found in Western cuisine. It can be lightly steamed and eaten plain or with a sauce as a side dish, and incorporated in casseroles and stir-fries. Sprouting vegetables such as broccoli and cauliflower should never be overcooked.

CABBAGE: As well as the more common red and green cabbages shown, the loose-headed savoy is worth trying. Cabbage should be steamed or lightly pan-fried, rather than boiled, and retain some crispness when cooked. To keep its colour, red cabbage needs to be cooked with an acidic ingredient such as apple or wine. Raw green cabbage is used to make coleslaw.

CAPSICUM (Pepper): Capsicum are readily available in red and green forms, sometimes yellow, and occasionally in a dark purple that is almost black. In Britain they are known as sweet peppers, as indeed they are. Capsicum are delicious raw, and can be cut into decorative rounds or neat strips to dress up a salad. Cooked, they add a distinctive flavour to tomato-based casseroles. Whole capsicum can be stuffed with a meat or rice mixture and baked; red capsicum are superb grilled, peeled and used in salads, savoury tarts or antipasto platters.

CARROTS: Sweet, crunchy carrots must be one of the most popular vegetables, particularly with children. A root vegetable, they can be served steamed, caramelized, and as a side dish, used in all manner of cooked dishes, or cut into matchstick strips as a salad ingredient or as part of a platter of crudités. Baby or Dutch carrots can simply be washed and eaten whole, cooked or raw. Leave the leafy top intact and serve with a dip at parties or add to school lunchboxes. Carrots are loaded with vitamin A and fibre.

CAULIFLOWER: A vegetable with a long history, first introduced into Europe in the 13th century, the cauliflower consists of a head of tightly packed white florets encased in tough leaves. The florets are an attractive addition to salads (blanched or raw) or stir-fried dishes. Cauliflower should be cooked only until tender to retain vitamins and flavour. It is most popularly served with a creamy cheese sauce. Baby cauliflowers can be served whole, one per person.

CELERY: Most often eaten raw as a salad vegetable, sometimes in decorative curls, celery is also frequently used in Italian cookery and in poultry and pork stuffing. It is also delicious braised and served as a side dish, with a white sauce. Use the leafy top as a flavouring for cooked dishes.

CHILLI: From the capsicum family, chillies are spicy and very highly flavoured. The smaller and redder the chilli the hotter it is (such as the tiny bird's eye, pictured here). Chilli is used as a spice more than as a vegetable and is rarely ever eaten whole. Finely sliced or crushed into a paste it is added to Mexican, Indian and Asian food. Use chillies with caution until you decide how much heat you like in a particular dish. To reduce the heat of chillies, cut off the membranes and seeds and discard.

CHINESE VEGETABLES: Pictured are some of the many varieties readily available in Chinese grocery stores and supermarkets. Bok choy (pak choi) is a type of cabbage (short with white stems). Choy sum resembles spinach and has yellow flowers. Leafy vegetables add colour and warmth to braised dishes and combine well with meats, noodles and oyster, mushroom and black bean sauces. These vegetables are blanched and added to noodle soups at the last minute, finely chopped as a garnish or served as a side dish. Long snake beans (doh gok) are included in many Southeast Asian stir-fries, curries and soups.

CHIVES: This mild herb is a member of the onion and garlic family. Its subtle taste combines well and the small green flecks of chopped chives are attractive against the pale colours of scrambled eggs, mashed potato and white sauces. The pretty blue flowers are edible and can be used in salads.

CORN: Originally from South America, this vegetable is available both fresh and packaged. Corn on the cob and baby corn are fresh and plentiful in summer; when not available, use tinned creamed corn or canned or frozen corn niblets. Corn is versatile; use it in fritters, soup, chowder, or Asian dishes. Barbecue or steam whole cobs; serve hot with butter, salt and pepper.

CORIANDER (Cilantro): A leafy green plant with tubular roots, this herb adds a crisp, spicy freshness and full flavour to Thai dishes and to Indian and Mexican food (although some dislike its 'earthy' flavour). Use the leaves in salads and salsa. The roots are ground and added to curry paste.

CUCUMBER: Popular around the world for its cooling properties, fresh cucumber is combined with yoghurt in India to accompany curries, mixed with yoghurt and garlic in Greece as a dip, peeled and used in sandwiches in England, added to gazpacho (cold soup) in Spain. The smaller Lebanese (short) variety are crisper and have a better flavour than larger ones. Apple cucumbers are white, spherical and are excellent in salads.

DILL: A sweet-scented herb of the parsley family, with soft, thin feathery leaves, dill is often used as a garnish for fish such as salmon or trout, and goes well with egg dishes and sauces. It is used for pickling (hence dill pickles) and to make dill vinegar. Dill seeds, which have a stronger flavour than the leaves, are also available.

EGGPLANT (Aubergine): Eggplant is available all year round, in shapes ranging from globe, egg and sausage, and colours from white through to the more usual purple. It originated in India, and was quickly adopted by the Arabs. Today it is found in many cuisines. The classic dish made with eggplant is the Greek moussaka, but it is used as a vegetable in Indian and Thai curries, stuffed or baked on its own. Cold grilled (broiled) eggplant is also found in salads and on antipasto platters, and in Italian-style focaccia sandwiches. Baby eggplant may be pickled in vinegar with garlic and oregano and used like pickled gherkins.

FENNEL: A squat, bulbous plant which looks like a celery heart, fennel has feathery leaves which resemble dill. It has an anise flavour, and looks and tastes good in stews, specially those using fish, pork or veal. Fennel grows in abundance in the Mediterranean region, and is thus used widely in Italian and other dishes of the area. Fennel can also be sliced into thin strips and used raw in salads, or cut in half lengthways, steamed and served with a white or cheese sauce.

GARLIC: Although there are several different types—red-skinned, violet and the large elephant garlic, the white garlic is the one most often seen. Used with discretion, it adds delicious flavour to cooked dishes, and raw it enhances salad dressings. If a milder flavour is preferred, whole cloves can be browned, then removed before the rest of the cooking.

GINGER: One of the oldest and most commonly used spices, particularly in Asian cooking, ginger comes in two forms. Raw or 'green' ginger is the root in its natural form; it is peeled and chopped to flavour stir-fries and curries. Fresh chopped green ginger is available in jars. Dried, ground ginger is mainly used in baking.

LETTUCE: No longer is lettuce synonymous with the traditional iceberg variety (lower left). Pictured are just some of those available (clockwise from top left): mignonette, cos (romaine), red

coral and salad mix (also known as mesclun). Lettuces fall into two general types—those with a firm, tightly packed head, and loose-leaf lettuce, with no heart. The former are perfect used as cups for salads and Asian stir-fries, while the latter can be picked a few leaves at a time as needed. Salad mix is a combination of a variety of baby lettuce leaves and edible flowers, sold by weight. A simple vinaigrette dressing will turn a bowl of mixed lettuce leaves into a special salad, but lettuce also lends itself to brief cooking, and makes a delicious soup.

MUSHROOMS: The common cultivated mushroom can be bought at three stages: (from top) the tiny buttons, slightly large cap mushrooms, which can be stuffed, and the larger flat (or field) mushrooms, which have more flavour than the smaller ones. Oyster mushrooms (at bottom) are also cultivated. They are chewier than the common variety, and have a shellfish flavour. Note: Do not be tempted by wild mushrooms; some varieties are highly poisonous.

ONIONS: Invaluable as a seasoning in most cuisines, onions also stand in their own right as a vegetable. Brown onions (left) have the strongest flavour; they are excellent in soups and stews. White onions (centre) are milder, often served raw in salads and used for stir-fries or caramelized until golden-brown. Mostly eaten raw, Spanish or red onions (right) are from a different family, and have a mild nutty taste.

PARSLEY, CURLY: Chopped or in whole sprigs, this biennial herb is usually used as a garnish. It is also used in soups, stews and pasta dishes, marrying particularly well with carrots, celery, cabbage, cauliflower, eggplant (aubergine) and spinach, and providing a rich source of vitamins, especially A and D, and minerals.

PARSLEY, CONTINENTAL: Also known as flat-leaf (Italian) parsley, continental parsley is used interchangeably with curly parsley, depending on preference. It does not keep quite as long as its cousin, but some prefer its stronger flavour and claim it has a less bitter taste.

PARSNIPS: This vegetable is a close relative of the carrot, but sweeter. It is served baked with a roast dinner, puréed, and used in soups (add a chopped parsnip to pea and ham soup) and in casseroles. Americans glaze parsnips with brown sugar and fruit juice.

PEAS: Members of the large legume family, peas appear in many different guises. Pictured are (from top) snow peas (mangetout) and sugar snap peas, both of which are eaten whole, frozen peas, split peas and fresh garden peas. Snow peas as also known as mangetout, meaning 'eat everything', but you may like to nip off the top and 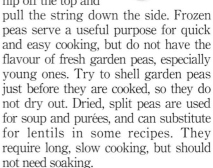 pull the string down the side. Frozen peas serve a useful purpose for quick and easy cooking, but do not have the flavour of fresh garden peas, especially young ones. Try to shell garden peas just before they are cooked, so they do not dry out. Dried, split peas are used for soup and purées, and can substitute for lentils in some recipes. They require long, slow cooking, but should not need soaking.

POTATOES: Although there are literally hundreds of varieties of potato grown throughout the world, you will commonly find only three or four at the greengrocer. Old potatoes (left) are floury and best for mashing. They can also be baked successfully, as can the red Pontiac. The Pontiac holds its shape well, and is good for sliced potato bakes. New potatoes (top right) can be boiled and used in potato salad, and baby chats just need steaming.

PUMPKIN: In America, the different varieties of pumpkin are known as winter squash to distinguish them from the soft-skinned summer squashes. They are all part of the gourd family. The traditional pumpkin (top), baked with the roast and potatoes for Sunday dinner, is Jarrahdale. But for a change, try the butternut (squash) or little golden nuggets. As well as baking, pumpkins are puréed, made into soup and used in pumpkin pie.

ROSEMARY: A herb with a pungent aroma somewhat like pine needles, rosemary should be used with care, lest it overpower the other ingredients. It is excellent with lamb and some mild-flavoured vegetables such as cabbage, squash and zucchini (courgette), and is one of the ingredients of a bouquet garni. Bunches of fresh rosemary can easily be hung up to dry and stored in an airtight container for future use.

SHALLOTS: Real shallots (at bottom) are small and brown, known in some countries as French shallots. In this book, both the bulbous and slender onions (pictured at top) are called spring onions (scallions), though in some countries they are erroneously called shallots. Leeks

stand as a vegetable on their own, as well as being used in soups, quiches and savoury tarts.

SPEARMINT: We call it mint, but the long-pointed leafed herb we most commonly use is spearmint, just one of the members of the mint family. Probably its best-known use is for mint sauce to accompany lamb, but it is also an important ingredient in Thai cooking, and adds flavour to boiled peas and potatoes. Mint leaves dry well in the microwave oven, keeping their colour and smell.

SPINACH: Although the two leafy greens pictured are both frequently described as spinach, they are from two different families. The large, bubble-textured leaves are *Beta vulgaris*, also known as silverbeet (Swiss

chard). At left is the true spinach, the small, smooth-leafed *Spinacea oleracea*, which is called English spinach. It has a much milder flavour, and as well as being cooked, makes an excellent salad.

SWEET POTATO: No relation to the potato, the sweet potato is similar in that it is a tuber. The three varieties shown differ mainly in their colouring (from left: red, white and orange). The orange sweet potato, from New Zealand, is sometimes called kumara. All are cooked in the same ways as potatoes, although sugar is sometimes used to emphasise their natural sweetness.

TOMATOES: Technically a fruit rather than a vegetable, tomatoes are, however, used exclusively in savoury dishes. Vine and Roma (plum) tomatoes are used interchangeably for cooking and salads—they are indispensable in Mediterranean cuisine. Little cherry tomatoes and the relatively new yellow pear variety are most commonly eaten raw, although they may be quickly stir-fried. The fashionable sun-dried tomatoes (at

bottom) may be rehydrated and used in place of fresh, but are more often found raw in salads and sandwiches. Bought tomatoes may need ripening for some days to develop full flavour.

TURNIPS: Like potatoes, turnips are a root vegetable, with variety in shape and colour, but all with a similar peppery taste. Baby turnips are relative newcomers to vegetable markets. They only need to be washed, topped and tailed to be used whole in dishes such as navarin of lamb or used raw in salads. Turnip tops may be washed and cooked in the same way as spinach.

ZUCCHINI (Courgette): In America, zucchini and the little yellow and green squash shown are known as summer squash, and zucchini are actually baby marrows. Young and tender, they need only to be steamed briefly and served with butter and a little black pepper. Zucchini combine well with tomatoes, and are used in ratatouille, and the flowers can also be stuffed with a savoury mixture, then dipped in batter and deep-fried in oil.

SNACKS

CHEESE AND OLIVE SLICE

Preparation time: 15 minutes
Total cooking time: 15 minutes
Makes about 20 pieces

1 red onion
30 g (1/4 cup) pitted black olives
1 red capsicum (pepper)
1 green capsicum (pepper)
2 tablespoons fresh basil leaves
3 teaspoons balsamic vinegar
2 garlic cloves, crushed
60 ml (1/4 cup) oil
3 garlic cloves, crushed, extra
1 large (30 x 40 cm/12 x 16 inch) piece of focaccia
100 g (3 1/2 oz) Cheddar cheese, grated
fresh basil leaves or other fresh herbs, to garnish

➤ PREHEAT OVEN to 180°C (350°F/ Gas 4). Line an oven tray with foil.
1 Slice onion and olives. Cut capsicum in halves. Remove seeds and membrane. Cut into fine strips. Finely shred basil leaves.
2 Combine capsicum, onion, olives, basil, vinegar and garlic in a medium bowl. Mix well. Cover and set aside.
3 Combine oil and extra garlic in a small bowl. Using a serrated knife, slice focaccia horizontally through centre. Brush focaccia halves with combined oil and garlic. Arrange the combined olive and capsicum filling evenly over the bottom half of focaccia. Sprinkle cheese over and top with remaining focaccia piece. Place on the prepared oven tray. Bake 15 minutes, or until the cheese melts. Cut into squares, garnish with basil leaves and serve hot or at room temperature.

COOK'S FILE

Storage time: Olive and capsicum filling can be made one day in advance. Store, covered, in the refrigerator. Assemble and cook squares just before serving.
Hints: Focaccia is flat Italian bread, available from most delicatessens, bakeries and some grocers.
To crush garlic, place the flat side of a large cook's knife over the peeled cloves and push down hard with the flat of your hand. Chop finely.
Variations: If focaccia is unavailable, use small bread rolls sliced in half, slices of large round continental loaves or sliced baguette.
Any of these ingredients can be added to the filling: sliced fresh mushrooms, marinated mushrooms, grilled eggplant (aubergine), sliced marinated artichoke hearts, sliced salami, ham or prosciutto, sliced turkey or chicken breast, small fresh prawns (shrimp) or sun-dried (sun-blushed) tomatoes.

FILO VEGETABLE POUCHES

Preparation time: 45 minutes
Total cooking time: 35–40 minutes
Makes 12

8 sheets filo pastry
125 g (4¹/2 oz) butter, melted
80 g (¹/2 cup) sesame seeds

Filling
465 g (3 cups) grated carrot
2 large onions, finely chopped
1 tablespoon grated ginger
1 tablespoon finely chopped
    fresh coriander (cilantro)
230 g (8 oz) can water chestnuts,
    rinsed and sliced

1 tablespoon miso
3 tablespoons tahini paste

➤ PREHEAT OVEN to 180°C (350°F/ Gas 4). Brush two oven trays with melted butter or oil.

1 To make Filling: Combine carrot, onions, ginger, coriander and 250 ml (1 cup) water in large pan. Cover, cook over low heat 20 minutes. Uncover, cook a further 5 minutes, or until all liquid has evaporated. Remove from heat, cool slightly. Stir in water chestnuts, miso and tahini. Season with pepper.

2 Place one sheet of filo pastry on work surface. Brush lightly with butter. Top with another three pastry sheets, brushing between each layer. Cut filo into six even squares. Repeat the process with the remaining pastry giving 12 squares in total.

3 Divide the filling evenly between each square, placing the filling in the centre. Bring the edges together and pinch to form a pouch. Brush the lower portion of each pouch with butter, then press in the sesame seeds. Place on prepared trays and bake for 10–12 minutes, or until golden brown and crisp. Serve hot with sweet chilli sauce, if liked.

COOK'S FILE

Storage time: Cook just before serving. Assemble pouches up to one day ahead. Store in refrigerator.
Hint: Miso is a salty soya bean paste, available from Asian food stores and supermarkets.

SPINACH CROQUETTES WITH MINTED YOGHURT SAUCE

Preparation time: 50 minutes
Total cooking time: 25 minutes +
  1 hour refrigeration
Makes 18

330 g (1¹/2 cups) short-grain
  rice
250 g (9 oz) feta cheese,
  crumbled
25 g (¹/4 cup) grated Parmesan
  cheese
2 eggs, lightly beaten
1 garlic clove, crushed
2 teaspoons grated lemon zest
60 g (¹/2 cup) chopped spring
  onions (scallions)
250 g (9 oz) packet frozen
  spinach, drained, squeezed
  of excess moisture
1 tablespoon freshly chopped
  dill
200 g (2 cups) dry breadcrumbs
2 eggs, lightly beaten, extra
oil, for deep-frying

Yoghurt Sauce
200 g (7 oz) plain yoghurt
2 tablespoons chopped fresh mint
2 tablespoons lemon juice

➤ COOK RICE in a large pan of boiling water until just tender; drain, rinse under cold water, drain again.

1 Combine rice, cheeses, eggs, garlic, lemon, onions, spinach and dill in a large bowl. Using wet hands, divide the mixture into 18 portions. Roll each portion into even-sized sausage shapes. Place on tray. Refrigerate for 30 minutes.

2 Spread breadcrumbs on a sheet of greaseproof paper. Dip croquettes into extra beaten egg mixture. Coat with breadcrumbs; shake off excess. Refrigerate a further 30 minutes.

3 To make Yoghurt Sauce: Combine the yoghurt, mint, lemon juice, salt and pepper in a bowl. Mix well. Cover, refrigerate until needed.

4 Heat oil in a deep heavy-based pan. Gently lower batches of croquettes into moderately hot oil with tongs or slotted spoon. Cook over medium high heat 2–3 minutes, or until golden and crisp and cooked through. Drain on paper towels. Repeat with remaining croquettes. Serve croquettes hot or cold with Yoghurt Sauce.

COOK'S FILE

Storage time: Croquettes can be made up to two days in advance. Cook just before serving.
Variation: Use fresh, lightly steamed spinach in place of frozen spinach if preferred. Use the same gram weight.

CAULIFLOWER FRITTERS WITH TOMATO RELISH

Preparation time: 35 minutes
Total cooking time: 30 minutes
Serves 4–6

1 small cauliflower
90 g (1/2 cup) peasemeal (see
     Note)
30 g (1/4 cup) self-raising flour
1 teaspoon ground cumin
1/4 teaspoon bicarbonate of soda
1 egg
200 g (7 oz) plain yoghurt
vegetable oil, for deep-frying

Tomato Relish
2 tablespoons vegetable oil
1 onion, finely chopped
400 g (14 oz) tomatoes, peeled
     and chopped
125 ml (1/2 cup) white wine
     vinegar
185 g (3/4 cup) sugar
1 garlic clove, crushed
1 teaspoon ground cumin
80 g (1/2 cup) sultanas
35 g (3/4 cup) finely chopped
     fresh coriander (cilantro)

➤ CUT CAULIFLOWER into large florets. Remove as much of the stem as possible without breaking florets. Wash and drain well. Pat dry with paper towels.

1 Combine peasemeal, flour, cumin and soda in a mixing bowl and make a well in the centre. Beat together the egg, yoghurt and 170 ml (2/3 cup) water. Pour onto the dry ingredients. Using a wooden spoon, stir until batter is smooth and free of lumps. Leave for 10 minutes.

2 **To make Tomato Relish:** Place the oil, onion, tomato, vinegar, sugar, garlic, cumin and sultanas in a pan.

Cover, cook over medium heat for 10 minutes. Bring to boil, reduce heat and simmer, uncovered, 5 minutes, or until mixture thickens and darkens slightly. Remove from heat. Stir in the coriander.

3 Heat oil in deep heavy-based pan. Dip florets in batter, drain off excess. Using a metal spoon or tongs, gently lower cauliflower into hot oil in small batches. Cook until golden brown, 3–5 minutes. Lift out with a slotted spoon and drain on paper towels. Serve hot with Tomato Relish.

THAI CORN PANCAKES WITH CORIANDER MAYONNAISE

Preparation time: 15 minutes
Total cooking time: 5 minutes each
batch
Serves 6

2 garlic cloves
1 small red chilli
2 cm (³/4 inch) piece fresh ginger
440 g (15¹/2 oz) can sweet corn
    kernels, drained
2 eggs
30 g (¹/4 cup) cornflour
    (cornstarch)
2 tablespoons fresh coriander
    (cilantro) leaves
1 tablespoon sweet chilli sauce
1 tablespoon peanut oil

Coriander Mayonnaise
160 g (²/3 cup) whole egg
    mayonnaise
60 ml (¹/4 cup) lime juice
10 g (¹/3 cup) coriander
    (cilantro) leaves, chopped
8 spring onions (scallions),
    finely chopped

➤ ROUGHLY CHOP the garlic; chop
the chilli and ginger.
1 Place half the sweet corn, eggs,
cornflour, coriander, garlic, chilli, ginger and chilli sauce in a food processor. Season with pepper. Using the
pulse action, process for 30 seconds,
or until smooth. Transfer to a bowl
and fold in the remaining corn.
2 Heat the oil in a large frying pan.
Spoon 2 tablespoons of corn mixture
into the pan and cook over a medium
heat for 2–3 minutes, or until golden.

Turn over and cook the other side for
1–2 minutes, or until cooked through.
Repeat process until all mixture is
used. Drain pancakes on paper towels.
3 To make Coriander Mayonnaise:
Combine the mayonnaise, lime juice,
coriander and spring onions in bowl.
Mix well. Add pepper to taste. Serve
pancakes hot or cool with a dollop of
Coriander Mayonnaise.

COOK'S FILE

Storage time: Coriander Mayonnaise
can be made up to one day in
advance. Store, covered, in refrigerator. Corn pancakes can be made several hours in advance. Reheat gently
just before serving.
Variation: Frozen corn kernels can
be used in place of canned corn.

SWEET POTATO MUFFINS

Preparation time: 15 minutes
Total cooking time: 25 minutes +
   10 minutes cooling
Makes 12

175 g (6 oz) sweet potato
250 g (2 cups) self-raising flour
125 g (1 cup) finely grated
   Cheddar cheese
90 g (3¹/4 oz) butter, melted and
   cooled
1 egg, lightly beaten
185 ml (³/4 cup) buttermilk

➤ PREHEAT OVEN to 180°C (350°F/
Gas 4). Brush melted butter or oil into
12 deep muffin tins.
1 Finely grate sweet potato. Sift flour
into large mixing bowl. Add sweet
potato and cheese, stir to combine.
Make a well in the centre.
2 Add butter, egg and buttermilk all
at once to dry ingredients. Using a
wooden spoon, stir until ingredients
are just combined; do not overbeat.
3 Spoon mixture into prepared muffin tins. Bake for 25 minutes, until
puffed and lightly golden. Turn onto a
wire rack to cool for 10 minutes before
serving. Serve warm with butter.

COOK'S FILE

Storage time: Cook just before serving. Muffins are best eaten on the day
they are made.

SAVOURY TARTS

Preparation time: 40 minutes
Total cooking time: 15–20 minutes
Makes 20

50 g (1³/4 oz) butter, melted
1 garlic clove, crushed
20 slices white bread

Filling
1 large carrot
1 large zucchini (courgette)
1 tablespoon oil
1 teaspoon grated fresh ginger
2 spring onions (scallions),
   finely sliced
90 g (3¹/4 oz) cauliflower, cut in
   small florets
1 tablespoon wholegrain mustard
¹/2 teaspoon dried basil
2 tablespoons chopped fresh
   chives

➤ PREHEAT OVEN to 180°C (350°F/
Gas 4). Grease two 12-cup shallow
patty tins.
1 Cut carrot and zucchini into short
thin strips.
2 Combine the butter and garlic in
small bowl. Remove crusts from bread
with a sharp knife. Using a rolling
pin, flatten each slice of bread. Cut
bread slices into rounds, using a 5 cm
(2 inch) plain or fluted cutter. Brush
bread with butter mixture. Place
bread rounds into the prepared tins,
press firmly. Bake for 10 minutes, or
until golden and crisp. Transfer to
wire rack to cool.
3 To make Filling: Heat oil in
medium heavy-based frying pan. Add
ginger and spring onions, cook over
medium heat 1 minute. Add carrot,
zucchini and cauliflower. Cook a further 5 minutes, or until vegetables are
tender. Add mustard, basil and chives,
season to taste and stir until combined. Remove from heat. Spoon one
tablespoon of the vegetable filling into
each tart case. Sprinkle with extra
chopped chives. Serve warm or cold.

COOK'S FILE

Storage time: Tart cases can be prepared and cooked a day in advance.
Store in an airtight container. Prepare
the filling and fill the bread cases just
before serving.
Variation: Use wholemeal or wholegrain bread in place of white.

*Sweet Potato Muffins (top)
and Savoury Tarts.*

ZUCCHINI FINGERS

Preparation time: 40 minutes
Total cooking time: 25 minutes
Makes 18 fingers

1 packet puff pastry
1 egg, lightly beaten
250 g (9 oz) cream cheese, soft
1 tablespoon mayonnaise
60 g (2¼ oz) sun-dried (sun-
    blushed) tomatoes in oil,
    well drained and chopped
1 garlic clove, crushed
1 teaspoon dried basil
6 zucchini (courgettes)
60 ml (¼ cup) olive oil
1 tablespoon lemon juice
2 tablespoons black olive paste

➤ PREHEAT OVEN to 210°C (415°F/Gas 6–7). Brush two oven trays with melted butter or oil.

1 Cut three puff pastry sheets into 8 x 24 cm (3 x 9½ inch) strips. Cut remaining pastry into 1 x 24 cm (½ x 9½ inch) strips. Brush pastry strips with egg. Place one thin strip of pastry down each side of the larger strip to form an edge. Prick pastry well with a fork. Place on prepared trays. Bake 15 minutes, or until crisp and golden. Cool on wire rack.

Using electric beaters, beat cream cheese until light and creamy. Add mayonnaise, tomatoes, garlic and basil; beat until combined. Set aside.

2 Cut ends from zucchini. Cut zucchini in thin slices lengthways. Place on cold grill (broiler). Brush with olive oil. Cook under medium-high heat for 10 minutes, or until golden brown. Drain. Drizzle with lemon juice.

3 Spread pastry shells with olive paste. Top with cheese mixture and smooth the surface. Arrange overlapping slices of zucchini over the cheese. Cut each into six fingers to serve.

COOK'S FILE

Storage time: Cream cheese filling can be prepared a day in advance. Cover and refrigerate. Cook and assemble fingers up to two hours before serving.

Hint: Olive paste is available from delicatessens and some supermarkets. If unavailable, make your own using pitted black olives and a small amount of olive oil. Process until smooth.

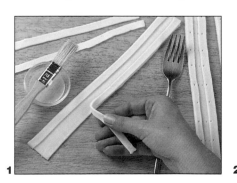

CRISP POTATO SKINS WITH CHILLI CHEESE DIP

Preparation time: 30 minutes
Total cooking time: 1 hour 15 minutes
Serves 4–6

6 medium potatoes
  (1.2 kg/2 lb 11 oz)
oil, for shallow-frying

Chilli Cheese Dip
1 tablespoon oil
1 small onion, finely chopped
1 garlic clove, crushed
1 teaspoon mild chilli powder
185 g (3/4 cup) sour cream
250 g (2 cups) grated Cheddar
  cheese

➤ PREHEAT OVEN to 210°C (415°F/ Gas 6–7).

1 Scrub potatoes and dry thoroughly; do not peel. Prick each potato twice with a fork. Bake for 1 hour, until skins are crisp and flesh is soft when pierced with a knife. Turn once during cooking. Remove from oven and cool.

2 Cut the potatoes in half and scoop out flesh, leaving about 5 mm (1/4 inch) of potato in the shell. Set aside flesh for another use. Cut each half into three wedges.

3 Heat oil in a medium heavy-based pan. Gently place batches of potato skins into moderately hot oil. Cook for 1–2 minutes, or until golden and crispy. Drain on paper towels. Serve immediately with Chilli Cheese Dip.

4 To make Chilli Cheese Dip: Heat oil in a small pan. Add onion and cook over a medium heat 2 minutes, or until soft. Add garlic and chilli powder, cook 1 minute, stirring. Add sour cream and stir until it is warm and thinned down slightly; add cheese and stir until melted and mixture is almost smooth. Serve hot.

COOK'S FILE

Storage time: Potatoes may be baked and prepared for frying up to eight hours in advance. Fry just before serving. Prepare Chilli Cheese Dip just before serving.

Variation: Cut the whole baked potatoes into wedges, fry and serve as described above.

1

2

3

4

SPINACH AND OLIVE BITES

Preparation time: 1 hour +
  1 hour refrigeration
Total cooking time: 15 minutes
Makes 30

250 g (2 cups) plain (all-purpose)
  flour
200 g (7 oz) butter, cut into
  7 mm (¼ inch) cubes

Filling
60 g (2¼ oz) English spinach
  leaves
100 g (3½ oz) feta cheese
2 tablespoons chopped pitted
  black olives
2 teaspoons chopped fresh
  rosemary
1 garlic clove, crushed
2 tablespoons pistachios
1 egg, lightly beaten

➤ SIFT FLOUR into a large mixing bowl; stir in the cubed butter until just combined.

1 Make a well in the centre of the flour, add almost all 185 ml (¾ cup) water. Mix to a slightly sticky dough with a knife, adding more water if necessary. Gather dough into a ball.

2 Turn onto a well-floured surface, and lightly press together until almost smooth. Do not overwork dough. Roll out to a neat 20 x 40 cm (8 x 16 inch) rectangle, trying to keep the corners fairly square. Fold the top third of the pastry down and fold the bottom third of the pastry up over it. Make a quarter turn to the right so that the edge of the top fold is on the right. Re-roll pastry to a 20 x 40 cm (8 x 16 inch) rectangle, and repeat folding step. Wrap pastry in plastic wrap and refrigerate for 30 minutes.

3 Repeat previous step, giving a roll, fold and turn twice more. Refrigerate for 30 minutes. Folding and rolling gives the pastry its flaky characteristics. Roll out pastry on a well-floured surface to a 3 mm (⅛ inch) thickness; cut out thirty 8 cm (3 inch) rounds.

4 Preheat oven to 180°C (350°F/Gas 4). Brush a large baking tray with melted butter or oil. Wash and dry spinach thoroughly, shred finely and place in mixing bowl. Crumble the feta on top, add olives, rosemary and garlic.

5 Spread pistachios on a baking tray and toast under a moderately hot grill (broiler) for 1–2 minutes. Cool and chop finely. Add to spinach mixture with egg, stir until well combined.

6 Place 2 teaspoonfuls of mixture in the centre of each round, fold in half and pinch edges to seal. Place on prepared tray, brush lightly with beaten egg and bake for 15 minutes, until golden and crisp. Serve hot.

COOK'S FILE

Storage time: Flaky pastry can be made up to one day in advance. Store in refrigerator. Assemble bites and cook just before serving.

Hints: Homemade flaky pastry is delicious and worth the effort if time permits. Have all the ingredients, equipment and room temperature as cool as possible—if it's too warm, the butter in the pastry will melt, making it difficult to work with.

Variation: Frozen puff pastry may be substituted in this recipe. Use ready-rolled sheets of butter puff pastry for best results.

Note: Feta cheese is available from delicatessens and supermarkets. It is a dry white cheese, cured in brine. The Bulgarian and Greek varieties are strongest in flavour. Originally made from goat's or ewe's milk, nowadays it is often made from cow's milk.

1

2

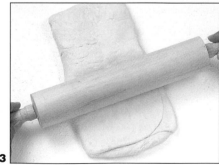

3

4

5

6

VEGETABLE SAMOSAS

Preparation time: 20 minutes
Total cooking time: 30 minutes
Makes 32

1 tablespoon ghee or oil
1 small onion, finely chopped
1 garlic clove, crushed
2 teaspoons grated fresh ginger
1 teaspoon mustard seeds
1 teaspoon ground cumin
1/2 teaspoon turmeric
1/4 teaspoon chilli powder
300 g (10 1/2 oz) potatoes
80 g (1/2 cup) frozen peas
2 tablespoons chopped fresh
    coriander (cilantro)
4 sheets frozen shortcrust
    pastry
oil, for deep frying

Yoghurt Dip
1/2 small cucumber
250 g (1 cup) plain yoghurt
2 tablespoons finely chopped
    fresh mint

➤ HEAT GHEE or oil in medium pan,
add onion, garlic and ginger. Cook over
low heat for 5 minutes, or until onion is
soft. Add spices, cook for 1 minute.

1 Cut potatoes into 7 mm (1/4 inch)
cubes. Add to the pan and stir to combine. Add 185 ml (3/4 cup) water, cover
and cook, stirring occasionally, for
5–10 minutes, or until potatoes are
just tender. Drain.

2 Remove the pan from heat and stir
in peas, coriander and salt; cool.

3 Cut sixteen 12 cm (5 inch) pastry circles and cut each in half. Fold each
semicircle in half; pinch straight sides
together to form cones. Spoon 2 teaspoons of filling into each cone. Pinch
edge to seal. Heat oil in a medium pan.
Cook samosas in batches in moderately
hot oil 2–3 minutes until crisp and gold-
en. Drain on paper towels. Serve warm
with Yoghurt Dip.

4 To make Yoghurt Dip: Peel
cucumber, remove seeds and finely
chop flesh. Combine with yoghurt and
mint in a small bowl.

COOK'S FILE

Storage time: The vegetable filling
can be made one day ahead. Assemble
and cook just before serving.

1

2

3

4

AVOCADO SALSA

Preparation time: 15 minutes
Total cooking time: 1 minute
Serves 6

1 red onion
2 large avocados
¼ teaspoon lime juice
1 tomato
1 small red capsicum (pepper)
1 teaspoon ground coriander
1 teaspoon ground cumin
15 g (¼ cup) chopped fresh
    coriander (cilantro) leaves
2 tablespoons olive oil
4–5 drops Tabasco

➤ FINELY CHOP the red onion.

1 Cut avocados in half, remove stone and carefully peel. Finely chop flesh, place in a medium mixing bowl and toss lightly with lime juice.

2 Cut tomato in half horizontally and squeeze gently to remove seeds; chop finely. Remove seeds and membrane from capsicum, chop finely.

3 Place the ground coriander and cumin in a small pan, stir over medium heat 1 minute to enhance fragrance and flavour; cool. Add all ingredients to avocado in bowl and gently combine, so that the avocado retains its shape and is not mashed. Refrigerate until required. Serve at room temperature with corn chips.

COOK'S FILE

Storage time: This dish can be made up to four hours in advance.
Hints: For immediate use, choose an avocado that will just give to a gentle squeeze. Avocados should have skin without blemishes or brown patches. Green-skinned varieties should have shiny skins. Overripe avocados are sometimes offered at a low price, but on opening can be rancid and brown. Better still, buy avocados that are just underripe, and check them daily to use at their peak. Storing avocados with bananas will hasten the ripening process. Place avocados in a brown paper bag on top of the refrigerator to ripen them overnight.

QUICK ONE-STEP RECIPES

Simple, time-saving methods that bring out the best in vegetables. These recipes make both hot and cold dishes with the minimum of fuss and bother—just what the modern cook ordered. All recipes make four servings.

Carrots

HONEY-GLAZED CARROTS

Peel and cut 2 carrots into thin diagonal slices. Steam or microwave until just tender. Do not overcook. Drain excess liquid from pan. Add 20 g (1/2 oz) butter and 1–2 teaspoons honey. Toss until well combined and the butter has melted. Sprinkle with chopped fresh chives or chopped fresh parsley. Serve hot.

CARROT RIBBONS

Trim the ends from 3 carrots. Peel carrots lengthways into strips using a sharp vegetable peeler. Heat 20 g (1/2 oz) butter in medium heavy-based frying pan. Add 1 teaspoon soft brown sugar and the carrot ribbons. Toss until well coated with butter, and carrots are tender but still crisp. Do not overcook. Add 1 tablespoon freshly chopped coriander (cilantro) leaves. Drizzle sparingly with balsamic vinegar if desired. Serve hot.

HERBED CARROT STICKS

Peel and cut 3 carrots into thin matchstick lengths. Steam or microwave until just tender. Do not overcook. Add 20 g (1/2 oz) butter and 1 tablespoon finely chopped fresh herbs or 2 teaspoons dried mixed herbs. Toss until well coated and butter has melted. Season to taste with salt and freshly ground black pepper. Serve hot.

GARLIC BUTTERED BABY CARROTS

Trim 12 baby (Dutch) carrots. Steam or microwave until just tender. Do not overcook. Heat 30 g (1 oz) butter in a medium pan. Add 2 crushed garlic cloves, 1/2 teaspoon sugar and 1 teaspoon finely grated lemon rind. Cook for 1 minute. Add carrots, stir until well coated and heated through. Sprinkle with finely chopped fresh herbs to taste. Serve hot.

Clockwise from top: Garlic Buttered Baby Carrots, Herbed Carrot Sticks, Carrot Ribbons, Honey-glazed Carrots.

POTATO CHIPS OR CURLS

GOLDEN ROASTED POTATOES

Potatoes

CREAMY POTATO

POTATO CHIPS OR CURLS

Peel 4 potatoes. Cut lengthways in 1 cm- (1/2 inch) thick slices, then into 1 cm- (1/2 inch) wide sticks or peel long potato strips with a vegetable peeler. Cook in a deep pan of moderately hot oil for 4–5 minutes for chips, 2–3 minutes for curls, or until golden and crisp. Drain on paper towels. Serve hot.

GOLDEN ROASTED POTATOES

Peel 4 potatoes. Place in baking dish, brush liberally with combined 1 tablespoon olive oil and 20 g (1/2 oz) melted butter. Bake at 210°C (415°F/Gas 6–7) for 20 minutes, brush with oil mixture. Bake further 30 minutes, or until crisp and golden. Serve hot.

CREAMY POTATO

Peel and chop 4 potatoes. Cook in a large pan of boiling water until just tender; drain and mash. Add 20 g (1/2 oz) butter and 2–3 tablespoons cream. Season. Stir until smooth and creamy. Sprinkle with chopped fresh herbs. Serve hot.

DUCHESS POTATOES

Peel and chop 4 potatoes. Cook in a large pan of boiling water until just tender; drain and mash. Add 3 egg yolks, 2 tablespoons cream and 2 tablespoons grated Parmesan cheese. Mix thoroughly. Pipe mixture into swirls onto greased oven trays. Bake in 210°C (415°F/Gas 6–7) oven for 20 minutes, or until golden brown. Sprinkle with paprika. Serve hot.

DUCHESS POTATOES

QUICK CHEESY POTATO BAKE

Peel and thinly slice 4 potatoes. Thinly slice 1 onion. Layer potato and onion slices in an ovenproof baking dish. Sprinkle grated Cheddar cheese between each layer. Pour over combined 125 ml (1/2 cup) cream, 185 ml (3/4 cup) milk and 1 teaspoon mustard powder. Sprinkle top with extra grated cheese and chopped chives. Bake at 180°C (350°F/Gas 4) for 40 minutes, or until cooked. Serve hot.

HERBED NEW POTATOES

Wash 12 small new potatoes. Cook in a large pan of boiling water until tender; drain in a colander. Toss with 60 g (2 1/4 oz) melted butter and 2 teaspoons each of freshly chopped basil and chives or parsley. Serve hot.

POTATO SALAD

Peel and chop 4 potatoes into 1.5 cm (5/8 inch) cubes. Cook in a pan of boiling water until just tender; drain. Rinse and drain again; leave to cool. Combine 60 g (1/4 cup) whole egg mayonnaise, 2 chopped spring onions (scallions), 1 chopped stalk celery, 2 teaspoons lemon juice and 1 rasher bacon, cooked and finely chopped. Mix well. Add potato. Toss through. Add extra chopped herbs if desired. Cover and refrigerate. Serve cold.

HASSELBACK POTATOES

Peel and halve 4 potatoes. Place potatoes cut-side down. Use a sharp knife to make thin slices in potatoes, taking care not to cut right through. Place potatoes cut-side up in baking dish. Brush with 1 tablespoon olive oil combined with 20 g (1/2 oz) melted butter. Sprinkle with lemon pepper. Bake in 210°C (415°F/Gas 6–7) oven for 45 minutes, or until golden and slightly crisp. Serve immediately.

QUICK CHEESY POTATO BAKE

HERBED NEW POTATOES

HASSELBACK POTATOES

POTATO SALAD

31

Broccoli

BROCCOLI WITH CHEESE SAUCE

Cut 250 g (9 oz) broccoli into small florets. Steam until just tender. Heat 40 g (1½ oz) butter in pan; add 2 tablespoons plain (all-purpose) flour. Stir over low heat 2 minutes, or until lightly golden and bubbling. Gradually add 250 ml (1 cup) milk, stirring until smooth. Stir constantly over medium heat until mixture boils and thickens. Take off heat; stir through 40 g (⅓ cup) grated Cheddar cheese until melted. Pour over hot broccoli to serve.

BROCCOLI WITH CASHEWS

Cut 250 g (9 oz) broccoli into small florets. Heat 1 tablespoon olive oil in heavy-based frying pan. Add 1 clove crushed garlic and 80 g (½ cup) cashew nuts (unsalted). Stir over medium heat 2 minutes, or until lightly golden. Add broccoli, stir-fry for 3–4 minutes, or until just tender. Serve hot.

BROCCOLI WITH BACON AND PINE NUTS

Cut 250 g (9 oz) broccoli into small florets. Heat 2 teaspoons oil in wok or heavy-based frying pan. Add 2 rashers thinly sliced bacon. Cook over medium heat 2 minutes. Add broccoli, stir-fry 3–4 minutes, or until just tender. Stir in 2 tablespoons toasted pine nuts and 1 tablespoon chopped fresh chives. Serve hot.

BROCCOLI WITH CHEESE SAUCE

BROCCOLI WITH CASHEWS

BROCCOLI WITH BACON AND PINE NUTS

BROCCOLI AND ONION STIR-FRY

BROCCOLI AND ONION STIR-FRY

Cut 250 g (9 oz) broccoli into small florets. Slice an onion into 8 wedges. Heat 2 teaspoons sesame oil and 2 teaspoons vegetable oil in wok or frying pan. Add broccoli and onions, cook until just tender. Stir in 2 teaspoons soy sauce and 3 teaspoons sweet chilli sauce. Sprinkle with herbs. Serve.

BUTTERED BROCCOLI AND HERBS

Cut 250 g (9 oz) broccoli into small florets. Heat 40 g (1½ oz) butter in wok or heavy-based frying pan. Add broccoli. Cover and cook over medium heat until just tender. Stir through 15 g (¼ cup) chopped fresh mixed herbs. (Use any combination: basil, mint, chives, parsley, oregano, marjoram, thyme, coriander/cilantro or dill). Serve hot.

LEMON BROCCOLI

Cut 250 g (9 oz) broccoli into small florets. Steam broccoli until just tender. Toss broccoli in 2 teaspoons olive oil combined with 3 teaspoons lemon juice. Serve hot.

BROCCOLI AND MUSHROOMS

Cut 250 g (9 oz) broccoli into small florets. Heat 30 g (1 oz) butter and 1 crushed garlic clove in heavy-based frying pan. Add 4 sliced button mushrooms. Cook over medium heat 2 minutes, or until tender. Remove from pan, set aside. Add broccoli, stir-fry 3–4 minutes until tender. Return mushrooms to pan, stir until heated through. Serve hot.

BROCCOLI WITH MUSTARD BUTTER

Cut 250 g (9 oz) broccoli into medium florets. Steam broccoli until just tender. Combine 50 g (1¾ oz) softened butter, 2 teaspoons Dijon mustard and freshly ground black pepper to taste. Mix well. Serve over hot broccoli.

BUTTERED BROCCOLI
AND HERBS

EMON BROCCOLI

BROCCOLI AND
MUSHROOMS

BROCCOLI WITH
MUSTARD BUTTER

**CABBAGE AND SPRING
ONION STIR-FRY**

**SWEET RED CABBAGE
WITH CARAWAY SEEDS**

SAUTEED CABBAGE

QUICK COLESLAW

Cabbage

SWEET RED CABBAGE WITH CARAWAY SEEDS

Finely shred ½ small red cabbage. Heat 30 g (1 oz) butter, 1 teaspoon caraway seeds, 1 teaspoon balsamic vinegar and 1 teaspoon soft brown sugar in a frying pan. Add cabbage, cook, stirring, 2–3 minutes, or until just tender. Serve hot.

SAUTEED CABBAGE

Finely shred ½ small green cabbage. Heat 20 g (½ oz) butter, 1 clove crushed garlic and 2 slices finely shredded ham or bacon in large pan. Cook 1 minute, add cabbage. Cook, stirring, 2–3 minutes, or until cabbage is just tender. Serve hot.

CABBAGE AND SPRING ONION STIR-FRY

Finely shred ½ small green cabbage. Heat 2 teaspoons olive oil and 20 g (½ oz) butter in heavy-based frying pan or wok. Add cabbage and 2 finely sliced spring onions (scallions). Stir-fry 2–3 minutes, or until just tender. Serve hot.

QUICK COLESLAW

Finely shred ½ small green cabbage. Combine with 2 grated carrots, 1 stalk finely chopped celery, 1 finely chopped onion, 1 finely chopped small green or red capsicum (pepper) and 125 ml (½ cup) prepared coleslaw dressing in a large bowl. Toss well to combine. Add 7 g (¼ cup) freshly chopped mixed herbs if desired. Chill before serving.

CABBAGE AND BEANS

GARLIC PEPPER CABBAGE

CABBAGE AND POTATO CAKES

SWEET CHILLI CABBAGE

SWEET CHILLI CABBAGE

Finely shred 1/2 small Chinese or green cabbage. Heat 2 teaspoons sesame oil in frying pan or wok. Add 2–3 teaspoons sweet chilli sauce, 1 teaspoon soy sauce and cabbage. Stir-fry 2–3 minutes, or until just tender. Serve hot.

CABBAGE AND POTATO CAKES

Combine 1 cup cooked cabbage, 115 g (1/2 cup) roughly mashed potato, 1 finely chopped spring onion (scallion), 2 lightly beaten eggs and salt and freshly ground pepper to taste, and mix well. Heat oil or butter in frying pan. Cook spoonfuls of mixture in batches 2 minutes each side, or until golden. Drain on paper towels. Serve hot.

GARLIC PEPPER CABBAGE

Finely shred 1/2 small green cabbage. Heat 20 g (1/2 oz) butter and 1 teaspoon oil in heavy-based frying pan or wok. Add 1–2 teaspoons garlic pepper seasoning and the cabbage. Stir-fry 2–3 minutes, or until just tender. Serve hot.

CABBAGE AND BEANS

Finely shred 1/2 small green or red cabbage. Heat 1 tablespoon olive oil in frying pan or wok. Add 1 crushed garlic clove, 1/4 teaspoon sugar, 12 finely shredded green beans and the cabbage. Stir-fry 3–4 minutes, or until vegetables are just tender. Serve hot, sprinkled with cracked or freshly ground black pepper to taste.

Pumpkin

CANDIED PUMPKIN

Peel and cut 500 g (1 lb 2 oz) butternut pumpkin (squash) in thin slices. Lay slices overlapping in ovenproof dish. Heat 40 g (1½ oz) butter, 2 tablespoons cream and 1 tablespoon soft brown sugar in a pan on low heat. Stir until smooth. Pour over pumpkin. Bake at 180°C (350°F/Gas 4) for 35 minutes, or until tender. Serve sprinkled with chopped chives.

BAKED PUMPKIN

Peel 500 g (1 lb 2 oz) pumpkin. Cut into large pieces. Place in baking dish. Brush liberally with combined 20 g (½ oz) melted butter and 2 teaspoons olive oil. Bake at 180°C (350°F/Gas 4) for 40 minutes, or until dark golden and cooked.

PUMPKIN AND NUTMEG PUREE

Peel 500 g (1 lb 2 oz) pumpkin. Cut into pieces. Steam or microwave until soft. Mash with a fork or potato masher. Add ¼ teaspoon ground nutmeg and 20 g (½ oz) butter. Season. Stir until smooth. Spoon or pipe onto plate to serve.

PUMPKIN WITH CHIVE BUTTER

Peel 500 g (1 lb 2 oz) pumpkin. Cut into 2 cm (¾ inch) cubes. Steam or microwave until just tender. Combine 40 g (1½ oz) softened butter, 1 tablespoon chopped chives and freshly ground black pepper to taste. Serve on hot pumpkin.

CANDIED PUMPKIN

BAKED PUMPKIN

PUMPKIN AND NUTMEG PUREE

PUMPKIN WITH CHIVE BUTTER

**PUMPKIN WITH GARLIC
AND HERB BUTTER**

SWEET SPICED PUMPKIN

**FRIED PUMPKIN
RIBBONS**

PUMPKIN SOUP

PUMPKIN WITH GARLIC AND HERB BUTTER

Peel 500 g (1 lb 2 oz) pumpkin. Cut into thin slices. Steam or microwave until tender. Combine 40 g (1½ oz) softened butter, 1 crushed garlic clove, 2 teaspoons chopped fresh coriander (cilantro) leaves, 2 teaspoons chopped fresh mint. Mix well. Pipe rosettes onto paper-lined tray. Refrigerate until firm. Serve on hot pumpkin.

SWEET SPICED PUMPKIN

Peel 500 g (1 lb 2 oz) pumpkin. Cut into thin slices. Place on a greased foil-lined tray. Melt 40 g (1½ oz) butter. Brush over pumpkin. Combine ½ teaspoon each ground cumin, ground coriander, ground ginger and 1 teaspoon soft brown sugar. Mix well and sprinkle over pumpkin. Bake at 180°C (350°F/Gas 4) for 35 minutes, or until cooked. Serve hot.

PUMPKIN SOUP

Place 500 g (2 cups) of cooked, mashed pumpkin, 20 g (½ oz) butter, 125 ml (½ cup) chicken stock, 2 roughly chopped spring onions (scallions) and 60 ml (¼ cup) milk in food processor. Using the pulse action, process until smooth. Transfer to large pan, stir in 125 ml (½ cup) cream and season. Heat through. Serve hot with sour cream and a sprinkle of ground nutmeg.

FRIED PUMPKIN RIBBONS

Peel 500 g (1 lb 2 oz) pumpkin. Peel pumpkin into ribbons using a vegetable peeler. Heat a pan half filled with oil. Deep-fry pumpkin ribbons in batches until crisp and golden. Drain on paper towels. Sprinkle with salt and pepper. Serve hot.

Spinach

SHREDDED SPINACH AND BACON

Finely shred 1 bunch English spinach. Cut 2 rashers bacon into thin strips. Heat 2 teaspoons olive oil in frying pan. Add bacon, fry over medium-high heat until almost crisp. Add spinach, toss through until just wilted. Serve hot.

SPINACH WITH VINAIGRETTE

Finely shred 1 bunch English spinach. Combine 2 tablespoons olive oil, 2 teaspoons seeded mustard, 1 tablespoon balsamic vinegar, 1 crushed garlic clove, 1/2 teaspoon soft brown sugar and freshly ground black pepper to taste. Mix well, pour dressing over spinach, toss lightly. Serve cool.

SPINACH AND BUTTERED CHIVES

Finely shred 1 bunch English spinach. Place in a mixing bowl. Heat 20 g (1/2 oz) butter and 1 tablespoon oil in a small pan. Add 15 g (1/4 cup) finely chopped chives and 1 teaspoon cracked black pepper. Cook 1 minute. Add to spinach in bowl, mix well. Serve immediately.

CREAMED SPINACH

Tear 1 bunch English spinach into pieces. Heat 20 g (1/2 oz) butter in heavy-based frying pan. Add 1 finely sliced small onion. Cook 2–3 minutes, or until onion is soft. Add spinach, cook 1 minute. Stir in 60 ml (1/4 cup) cream, heat through. Sprinkle with nutmeg and grated Cheddar cheese. Serve hot.

SPINACH AND BUTTERED CHIVES

SHREDDED SPINACH AND BACON

SPINACH WITH VINAIGRETTE

CREAMED SPINACH

SWEET CHILLI SPINACH SALAD

Finely shred 1 bunch English spinach. Heat 3 teaspoons sesame oil, 2 teaspoons soy sauce, 1 tablespoon chilli sauce and 1 teaspoon fish sauce in a small pan. Cook 1 minute. Add 2 tablespoons chopped fresh coriander (cilantro) leaves. Toss through spinach to combine. Serve cool.

SWEET CHILLI SPINACH SALAD

SPINACH AND SPRING ONION SALAD

Tear 1 bunch English spinach leaves into pieces. Combine in bowl with 2–3 finely shredded spring onions (scallions). Season with salt and freshly ground black pepper to taste. Drizzle with 1 tablespoon olive oil combined with 1–2 tablespoons red wine or balsamic vinegar. Serve cool.

BASIL SPINACH SALAD

Tear 1 bunch English spinach into pieces. Combine with 15 g (1/4 cup) shredded basil leaves and 2 tablespoons toasted pine nuts. Finely slice 1 rasher bacon. Cook in small pan until crisp. Combine 2 tablespoons oil, 1 tablespoon white wine vinegar, 1 tablespoon sour cream, 1/2 teaspoon sugar and 1 clove crushed garlic. Mix well and drizzle over salad. Top with bacon and Parmesan cheese shavings. Serve cool.

EGGS FLORENTINE

Tear 2 bunches English spinach into pieces. Steam or microwave until tender. Combine spinach with 30 g (1 oz) butter, nutmeg and salt to taste. Divide mixture into 2 ramekin dishes. Top each with a poached egg. Sprinkle with grated Cheddar cheese. Bake at 180°C (350°F/Gas 4) for 5–10 minutes until cheese melts and browns. Serve hot.

BASIL SPINACH SALAD

SPINACH AND SPRING ONION SALAD

EGGS FLORENTINE

Beans

GARLIC AND BASIL BEANS

Trim tops from 20 green beans. Heat 1 tablespoon olive oil in frying pan or wok. Add 1 crushed garlic clove and beans. Cook, stirring, 2–3 minutes, or until beans are just tender. Stir in 1 tablespoon shredded fresh basil leaves. Serve hot.

PEPPER BEANS AND HAM

Top and tail 20 green beans. Heat 1 tablespoon olive oil in frying pan or wok. Add 2 slices finely sliced ham and 1 teaspoon cracked black pepper. Cook 1 minute. Add beans, stir-fry 2–3 minutes, or until beans are just tender. Serve hot.

BEANS HOLLANDAISE

Top and tail 20 green beans. Steam until just tender. Place 2 egg yolks in food processor bowl. Process 10 seconds, or until yolk is blended. With motor constantly running, add 150 g (5½ oz) melted butter in a thin stream until mixture is thick and creamy, then add 3 teaspoons lemon juice. Season with salt and freshly ground black pepper to taste. Spoon over warm beans. Serve immediately.

BEANS AND CASHEWS

Top and tail 20 green beans. Cut into 4 cm (1½ inch) diagonal lengths. Heat 2 teaspoons sesame oil in frying pan or wok. Add 1 crushed garlic clove and 50 g (⅓ cup) cashew nuts. Cook 2 minutes. Add beans, stir-fry 2–3 minutes, or until nuts are golden and beans just tender. Serve hot.

BEANS HOLLANDAISE

GARLIC AND BASIL BEANS

BEANS AND CASHEWS

PEPPER BEANS AND HAM

BEAN BUNDLES

BEAN BUNDLES

Top and tail 20 green beans. Divide beans into bundles of five. Tie together with a spring onion (scallion) green or chives. Steam or microwave until just tender. Sprinkle with lemon pepper. Serve hot.

BEAN AND WALNUT SALAD

Finely shred 20 green beans. Combine in a bowl with 7 g (1/4 cup) coriander (cilantro) leaves, 4 large red oak lettuce leaves torn in pieces, 1/4 finely shredded red capsicum (pepper), 35 g (1/3 cup) walnut halves, 2 tablespoons tarragon vinegar, 1–2 tablespoons peanut oil and 2 tablespoons chopped fresh mint. Mix well. Serve immediately.

BEANS IN HERB CREAM SAUCE

Top and tail 20 green beans. Heat 30 g (1 oz) butter in frying pan or wok. Add 1 crushed garlic clove and beans. Cook 2–3 minutes, or until just tender. Stir in 60 ml (1/4 cup) cream, 2 teaspoons chopped fresh rosemary, 1 tablespoon chopped fresh chives and 1 teaspoon chopped fresh thyme. Cook 1 minute more. Serve hot.

MINTED TOMATO AND BEANS

Cut 20 green beans in half diagonally. Heat 1 teaspoon oil in frying pan. Add 1 crushed garlic clove, 1 teaspoon grated fresh ginger, 1/2 teaspoon each of ground coriander, cumin and garam masala, and 2 chopped ripe tomatoes. Cook, stirring, for 1 minute. Add beans, cook 2–3 minutes, or until just tender. Stir in 1 tablespoon chopped mint. Serve hot.

BEAN AND WALNUT SALAD

BEANS IN HERB CREAM SAUCE

MINTED TOMATO AND BEANS

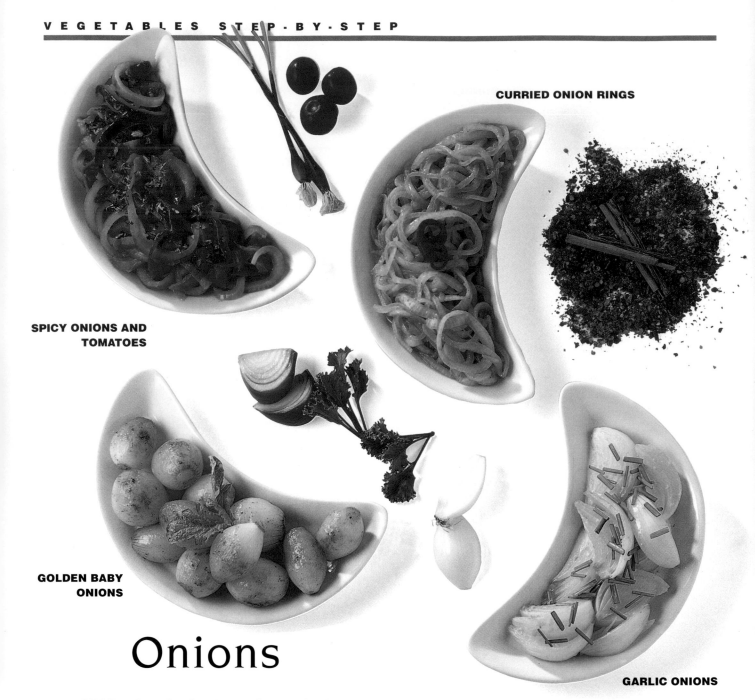

CURRIED ONION RINGS

SPICY ONIONS AND TOMATOES

GOLDEN BABY ONIONS

GARLIC ONIONS

Onions

SPICY ONIONS AND TOMATOES

Peel and thinly slice 2 medium red onions. Heat 20 g (1/2 oz) butter, 1 tablespoon oil, 1/2 teaspoon each ground cumin, coriander, turmeric and garam masala. Add onions, cook 2–3 minutes. Stir in 2 chopped medium ripe tomatoes. Cook 3 minutes more, or until onions are soft. Serve hot, sprinkled with chopped fresh coriander (cilantro).

GOLDEN BABY ONIONS

Peel 12 small onions, leaving bases intact. Heat 30 g (1 oz) butter, 1 tablespoon oil and 1/4 teaspoon ground sweet paprika in frying pan or wok. Add onions. Cook over medium heat 5 minutes, or until onions are golden brown and tender, then stir in 1/2 teaspoon soft brown sugar. Serve hot.

CURRIED ONION RINGS

Peel and cut 2 medium onions into thin rings. Heat 2 tablespoons olive oil in frying pan. Add 2 teaspoons curry powder and onions. Cook 5 minutes, or until onions are tender. Stir in 1/2 teaspoon soft brown sugar. Serve hot.

GARLIC ONIONS

Peel and cut 2 onions into eight wedges. Heat 20 g (1/2 oz) butter and 2 tablespoons oil in frying pan. Add 1–2 crushed garlic cloves and onions. Cook over medium heat 5–6 minutes, or until tender. Sprinkle with chopped chives. Serve hot.

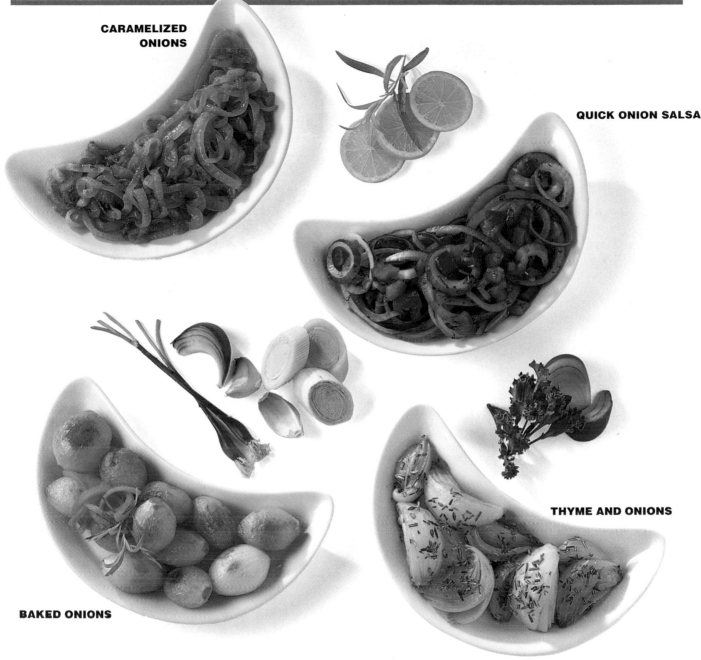

CARAMELIZED ONIONS

QUICK ONION SALSA

BAKED ONIONS

THYME AND ONIONS

CARAMELIZED ONIONS

Peel and cut 2 medium onions into thin rings. Heat 30 g (1 oz) butter and 1 tablespoon oil in heavy-based frying pan. Add onions, cook over low heat 10–12 minutes, or until onions are dark golden, stirring occasionally. Serve hot, drizzled with a little balsamic vinegar.

BAKED ONIONS

Peel 8 small onions, leaving bases intact. Place in baking dish. Brush liberally with combined 20 g (1/2 oz) melted butter and 1 tablespoon oil. Bake at 180°C (350°F/Gas 4) for 30 minutes, or until golden brown. Serve hot.

QUICK ONION SALSA

Peel and finely slice 1 large red onion. Combine in a bowl with 2 tablespoons lime juice, 1 tablespoon olive oil, 1 teaspoon soft brown sugar, 1 tablespoon chopped fresh coriander (cilantro), 1 chopped tomato, and 1 finely chopped jalapeno chilli. Mix well. Season to taste. Cover and set aside at room temperature 10 minutes before serving.

THYME AND ONIONS

Peel and cut 2 onions into eight wedges. Heat 10 g (1/4 oz) butter and 2 tablespoons oil in heavy-based frying pan. Add onions, cook over medium heat 5 minutes, or until tender and golden. Stir in 1 teaspoon each chopped fresh thyme and rosemary. Cook 1 minute more. Drizzle with vinegar. Serve hot.

Cauliflower

SPICED CAULIFLOWER

CAULIFLOWER WITH BACON

Cut 400 g (14 oz) cauliflower into small florets. Steam or microwave until just tender. Heat 1 teaspoon oil in a medium pan. Add 2 finely shredded rashers bacon, cook until browned. Add cauliflower and 2 finely chopped spring onions (scallions), stir to combine. Serve hot.

CAULIFLOWER CHEESE

Cut 400 g (14 oz) cauliflower into florets. Steam until just tender. Heat 30 g (1 oz) butter in a pan. Add 1 tablespoon plain (all-purpose) flour and cook, stirring, 1 minute. Gradually add 185 ml (3/4 cup) milk. Stir until sauce boils and thickens. Take off heat and stir in 40 g (1/3 cup) grated Cheddar cheese until melted. Pour over cauliflower. Serve hot.

CAULIFLOWER AU GRATIN

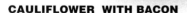

CAULIFLOWER WITH BACON

CAULIFLOWER CHEESE

SPICED CAULIFLOWER

Cut 400 g (14 oz) cauliflower into small florets. Heat 30 g (1 oz) butter and 1 tablespoon oil in wok or frying pan. Add 1/2 teaspoon each of ground turmeric, coriander, cumin and cinnamon. Cook 1 minute. Stir in 1 teaspoon soft brown sugar and cauliflower. Cook over medium heat until just tender. Serve hot with a dollop of yoghurt or sour cream.

CAULIFLOWER AU GRATIN

Cut 400 g (14 oz) cauliflower into florets. Steam or microwave until just tender. Heat 30 g (1 oz) butter in a pan. Add 2 teaspoons plain (all-purpose) flour and cook, stirring, 1 minute. Gradually add 125 ml (1/2 cup) milk. Stir until sauce boils and thickens. Season with salt, white pepper and nutmeg. Pour over cauliflower in an ovenproof dish. Sprinkle with 40 g (1/3 cup) grated cheese combined with 2 tablespoons breadcrumbs. Cook under hot grill (broiler) 2 minutes until golden.

CAULIFLOWER WITH TOMATO SAUCE

Cut 400 g (14 oz) cauliflower into medium florets. Steam or microwave cauliflower until just tender. Heat 1 tablespoon oil in medium pan. Add ½ teaspoon cracked pepper, 1 teaspoon Italian mixed herbs and 1 crushed garlic clove. Cook 1 minute. Add 440 g (15½ oz) can tomatoes, crushed. Bring to boil, reduce heat, simmer 5 minutes, or until reduced slightly. Pour sauce over and serve hot.

PARMESAN CAULIFLOWER

Cut 400 g (14 oz) cauliflower into small florets. Toss in combined 30 g (¼ cup) plain (all-purpose) flour, 2 tablespoons finely grated Parmesan cheese and 1 teaspoon dried mixed herbs. Heat 2 tablespoons oil and 40 g (1½ oz) butter in heavy-based frying pan. Gently cook cauliflower in batches until just tender. Drain on paper towels. Serve hot.

HOT CHILLI CAULIFLOWER

Cut 400 g (14 oz) cauliflower into small florets. Steam or microwave until just tender. Combine 40 g (1½ oz) melted butter, 1 tablespoon tomato paste (purée), 2 tablespoons chopped fresh coriander (cilantro) and ¼ teaspoon chilli powder (or to taste). Toss through cauliflower. Serve hot.

CAULIFLOWER WITH LIME BUTTER

Cut 400 g (14 oz) cauliflower into florets. Steam or microwave cauliflower until just tender. Combine 50 g (1¾ oz) softened butter, 1 tablespoon lime juice, 1 teaspoon finely grated lime zest, 1 crushed garlic clove and 1 teaspoon soft brown sugar. Mix well. Toss through cauliflower. Serve hot.

CAULIFLOWER WITH TOMATO SAUCE

HOT CHILLI CAULIFLOWER

PARMESAN CAULIFLOWER

CAULIFLOWER WITH LIME BUTTER

45

CREAMED PEAS

MINTED PEAS

SAUTEED PEAS AND SPRING ONIONS

SWEET CORIANDER PEAS

Peas

MINTED PEAS

Steam, microwave or lightly boil 310 g (2 cups) frozen green peas. Toss through 1 tablespoon finely chopped fresh mint and 1/4 teaspoon soft brown sugar. Serve hot.

CREAMED PEAS

Steam, microwave or lightly boil 310 g (2 cups) frozen green peas. Drain. Mash with a fork. Add 10 g (1/4 oz) butter and season to taste. Stir to combine. Serve hot.

SAUTEED PEAS AND SPRING ONIONS

Heat 30 g (1 oz) butter in frying pan. Add 310 g (2 cups) frozen green peas, 1 crushed garlic clove and 2 sliced spring onions (scallions). Stir for 2–3 minutes, or until tender. Serve hot.

SWEET CORIANDER PEAS

Heat 30 g (1 oz) butter in medium pan. Add 1 1/2 teaspoons lemon juice, 1/2 teaspoon sugar and 310 g (2 cups) frozen green peas. Cook over medium heat 2–3 minutes, or until just tender. Toss through 2 tablespoons finely chopped fresh coriander (cilantro) leaves. Serve hot.

PEAS AND BACON

PEAS AND BACON

Heat 2 teaspoons oil in a heavy-based frying pan. Add 2 finely chopped bacon rashers. Cook 1–2 minutes. Add 310 g (2 cups) frozen green peas, cook over medium heat 2–3 minutes, or until tender, stirring occasionally. Toss through 1 tablespoon chopped fresh chives and 1 tablespoon chopped fresh lemon thyme. Serve hot.

PEAS WITH BASIL AND TOMATO

Heat 2 teaspoons oil in heavy-based frying pan. Add 1 clove crushed garlic and 125 g (1/2 cup) chopped canned tomatoes in juice. Cook for 1 minute. Add 310 g (2 cups) frozen green peas and 1–2 tablespoons finely shredded basil. Cook for 2–3 minutes, or until just tender. Serve hot.

PEAS WITH BASIL AND TOMATO

PEPPERED PEAS AND GARLIC

Heat 1 tablespoon oil in heavy-based frying pan. Add 2 crushed garlic cloves and 1 teaspoon cracked black pepper. Stir in 310 g (2 cups) frozen green peas and 1/2 teaspoon sugar. Cook over medium heat 2–3 minutes, or until tender. Drizzle with balsamic vinegar if desired. Serve hot.

GOLDEN ONIONS AND PEAS

Heat 1 tablespoon oil and 30 g (1 oz) butter in heavy-based frying pan. Add 1 medium onion, peeled and finely sliced. Cook over low heat 5 minutes, or until onions are golden brown. Add 310 g (2 cups) frozen green peas. Cook 2–3 minutes, or until just tender. Serve hot.

PEPPERED PEAS AND GARLIC

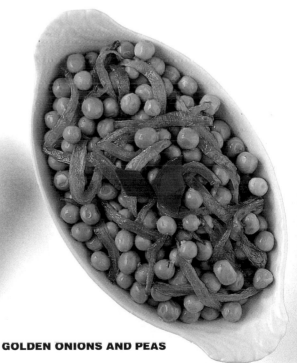

GOLDEN ONIONS AND PEAS

SOUPS & STARTERS

SALAD BASKETS WITH BERRY DRESSING

Preparation time: 20 minutes
Total cooking time: 15 minutes
Serves 4

12 sheets filo pastry
50 g (1³⁄₄ oz) butter, melted
160 g (5³⁄₄ oz) salad mix
8 cherry tomatoes, halved
1/2 small red capsicum (pepper),
   thinly sliced
1 small Lebanese (short)
   cucumber, thinly sliced
155 g (1 cup) blueberries, extra

Blueberry Dressing
125 ml (¹⁄₂ cup) olive oil
2 tablespoons balsamic vinegar
1 tablespoon soft brown sugar
50 g (¹⁄₃ cup) frozen blueberries,
   thawed, lightly crushed

➤ PREHEAT OVEN to 180°C (350°F/
Gas 4). Brush the outer base of four 8 cm
(3 inch) round, 125 ml (¹⁄₂-cup) ramekin
dishes with melted butter or oil. Line
two oven trays with baking paper.
1 Place one sheet of filo pastry on
work surface. Brush pastry lightly
with melted butter, place another
sheet on top. Repeat with a third
layer. Fold pastry in half. Using a
plate as a guide, cut a 21 cm (8¹⁄₂ inch)
circle out of the pastry with a sharp
knife. Place pastry over base of prepared ramekins. Carefully fold pastry
around the ramekin to form a basket
shape. Place on prepared trays. Repeat
process for three more baskets. Bake
15 minutes, or until golden. Carefully
remove filo basket from ramekin
while hot; place baskets on wire rack
to cool.
2 To make Blueberry Dressing:
Combine oil, vinegar and brown sugar
in small bowl, add crushed berries.
Mix well.
3 Combine salad mix, tomatoes, capsicum, cucumber and dressing in
bowl. Mix well. Spoon into pastry baskets, top with extra blueberries.

COOK'S FILE

Storage time: Filo baskets can be
prepared up to two hours ahead. Fill
with salad mixture just before serving. Dressing can be prepared one day
in advance. Store in an airtight container in refrigerator.
Hint: Salad mix (mesclun) is available
from most greengrocers. It includes a
wide variety of baby lettuce leaves
and edible flowers.

ROAST SWEET POTATO WITH CORIANDER PESTO AND SPRING SALAD

Preparation time: 30 minutes
Total cooking time: 35–40 minutes
Serves 6

850 g (1 lb 14 oz) orange sweet
    potato, peeled
2 tablespoons lemon juice
1 bunch asparagus
1 red capsicum (pepper)
1 small Lebanese (short)
    cucumber
160 g (5³/4 oz) salad mix

Coriander Pesto
100 g (²/3 cup) pine nuts
1 red chilli, seeded
2 garlic cloves
¹/2 bunch coriander (cilantro)
    leaves and roots, chopped
2 teaspoons lime juice
2 tablespoons oil

➤ PREHEAT OVEN to 180°C (350°F/
Gas 4).
1 Peel sweet potato and cut into 1 cm
(¹/2 inch) slices. Place in a bowl, cover
with iced water and lemon juice.
Leave for 5 minutes, drain and pat dry
with paper towels.
2 Heat oil in a deep baking dish on
top of stove. Add potato, lightly coat
with oil. Transfer dish to oven. Bake
30 minutes, or until golden. Remove;
keep warm.
3 To make Coriander Pesto:
Place pine nuts in small pan. Stir over
medium heat until golden. Remove
from heat, cool slightly. Place pine
nuts, chilli, garlic, coriander, juice and
oil in a food processor. Using the pulse
action, process for 30 seconds, or until
smooth. If mixture is too thick, add a
little extra oil to thin. Cover, set aside.
4 Plunge asparagus into a medium
pan of boiling water. Cook for
1 minute, or until just tender, drain.
Plunge into bowl of iced water, drain.
Pat dry with paper towels. Cut asparagus into 5 cm (2 inch) pieces. Cut capsicum into quarters. Remove seeds
and membrane. Place skin-side up on
grill (broiler) tray; brush with oil. Grill
(broil) 10 minutes, or until skin is black.
Cover with damp tea towel until cool.
Peel off skin. Cut capsicum into long
thin strips. Cut cucumber into matchstick thin strips. Arrange salad mix,
asparagus, capsicum and cucumber in
a serving bowl. Serve sweet potato
separately; top with Coriander Pesto.

COOK'S FILE

Storage time: Except for roast capsicum, which can be made one day
ahead, cook and assemble this dish
just before serving.
Variation: Ready-made roasted capsicum are available in supermarkets
and delicatessens.

HOT AVOCADO SALAD

Preparation time: 20 minutes
Total cooking time: 15 minutes
Serves 6

3 avocados
1 tomato
4 rashers bacon, optional
1 red onion, finely chopped
1 red capsicum (pepper), finely
    chopped
1 stalk celery, finely chopped
2 teaspoons sugar
2 teaspoons sweet chilli sauce
2 tablespoons balsamic vinegar
125 g (1 cup) grated Cheddar
    cheese

➤ PREHEAT OVEN to 180°C (350°F/
Gas 4). Brush a 25 cm (10 inch) pie
plate with melted butter or oil.
1 Cut avocados in half lengthways and
remove stones. Scoop out two-thirds of
flesh, roughly chop. Retain shells.
2 Peel, seed and finely chop tomato.
Trim bacon and place on cold grill
(broiler) tray. Cook under medium-
high heat until crisp. Let cool slightly
and chop finely. Combine avocado,
tomato, bacon, onion, capsicum and
celery in a bowl. Combine sugar, chilli
sauce and vinegar in a small screwtop
jar and shake well. Pour over ingredients in bowl and mix well.
3 Spoon filling into avocado halves,
sprinkle with cheese. Place in prepared dish. Bake 7–10 minutes, or

until heated through. Serve immediately with corn chips, crackers or thin
slices of white toast.

COOK'S FILE

Storage time: Cook this dish just
before serving.
Variation: This appetizer can also be
served as a Chilled Avocado Salad.
Add 2 tablespoons of olive oil to the
dressing. Replace the Cheddar cheese
with shavings of fresh Parmesan
cheese, cover filled avocados with
plastic wrap and refrigerate until
chilled before serving. Do not bake.
Note: Balsamic vinegar is flavourful,
aged wine vinegar from Modena,
Italy. It is available from supermarkets and delicatessens.

1

2

3

SWEET POTATO SOUP

Preparation time: 20 minutes
Total cooking time: 1 hour
Serves 8

1 kg (2 lb 4 oz) orange sweet
   potato
2 large onions
75 g (2½ oz) butter
1 garlic clove, crushed
1 tablespoon ground cumin
2 litres (8 cups) chicken stock
310 g (1¼ cups) crunchy peanut
   butter

1 tablespoon chilli sauce
40 g (¼ cup) chopped peanuts
15 g (¼ cup) chopped chives

➤ PEEL SWEET potato. Peel and chop onions.

1 Chop potato into 5 mm (¼ inch) cubes. Heat butter in a deep heavy-based pan. Add onions, cook over medium-high heat 10 minutes, or until golden brown.

2 Add garlic and cumin, stir-fry for 30 seconds. Add sweet potato, stir until well coated. Cover, reduce heat, cook 10 minutes. Shake pan occasionally to prevent sticking. Add stock,

bring to boil, then reduce heat and simmer for 10 minutes.

3 Stir in peanut butter and chilli sauce. Simmer gently, uncovered, for 30 minutes, stirring occasionally. Add salt to taste. Serve immediately, sprinkled with chopped peanuts and chives.

COOK'S FILE

Storage time: This soup can be prepared up to three days in advance. Store, covered, in refrigerator.
Hint: Spread sliced French sticks with butter creamed with curry powder to taste. Bake in moderate oven until golden. Serve hot with soup.

SPINACH AND SALMON TERRINE

Preparation time: 35 minutes
Total cooking time: 10 minutes
Serves 8

2 packets frozen chopped spinach
30 g (1 oz) butter
60 g (1/2 cup) chopped spring
    onions (scallions)
1 tablespoon chopped fresh dill
nutmeg, to taste
6 eggs
1 tablespoon cornflour
    (cornstarch)
1 tablespoon lime juice
25 g (1/4 cup) grated Parmesan
    cheese

Filling
30 g (1/4 cup) chopped spring
    onions (scallions)
1 teaspoon chopped fresh dill
1 tablespoon lime juice
1 tablespoon horseradish
250 g (9 oz) neufchatel cheese

200 g (7 oz) sliced smoked
salmon

➤ PREHEAT OVEN to 180°C (350°F/
Gas 4). Grease a shallow 30 x 25 cm
(12 x 10 inch) swiss roll tin. Line base
and sides with baking paper, extend
ing 5 cm (2 inch) extra at ends.
1 Thaw spinach and squeeze out
excess moisture. Heat butter in pan.
Add onions and dill, stir over medium
heat 1 minute. Add spinach, heat
through. Season with nutmeg and
pepper. Remove from heat.
2 Beat eggs in bowl. Blend cornflour
and juice in small bowl until smooth.
Combine with eggs and spinach mixture. Pour into tin. Bake 7 minutes, or
until firm to touch. Turn onto a damp
tea towel covered with sheet of baking
paper and sprinkled with Parmesan.
Cover with a cloth and leave to cool.
3 To make Filling: Process onions,
dill, juice, horseradish and cheese in
food processor 30 seconds until smooth.
4 Cut spinach base into three strips
(8 x 32 cm/3 x 13 inch). Place one strip
on board. Spread with filling, top with

a third of salmon slices. Spread sec-
ond strip with a thin layer of filling.
Place cheese-side down on salmon.
Repeat procedure with next layer.
Decorate with rolled smoked salmon
slices. Cut terrine into 2 cm (3/4 inch)
slices to serve. Serve as entrée or
brunch with a light salad.

COOK'S FILE

Note: Neufchatel (a type of cream
cheese) is available from delicatessens.

GRILLED TOMATOES WITH BRUSCHETTA

Preparation time: 15 minutes
Total cooking time: 35 minutes
Serves 4

1 loaf Italian bread
4 large ripe tomatoes
1/2 teaspoon dried marjoram
    leaves
2 tablespoons olive oil
2 tablespoons red wine vinegar
1 teaspoon soft brown sugar
2 tablespoons olive oil, extra
1 garlic clove, cut in half

110 g (1/2 cup) chopped
    marinated artichokes,
    undrained
1 tablespoon finely chopped
    flat-leaf (Italian) parsley

➤ CUT BREAD in thick slices. Preheat grill (broiler).
1 Cut tomatoes in half; gently squeeze out seeds. Place tomatoes cut-side up in shallow ovenproof dish. Place marjoram, oil, vinegar and sugar in a small screwtop jar. Season with salt and freshly ground black pepper and shake well. Pour the dressing over the tomatoes.
2 Cook the tomatoes under the hot

grill for 30 minutes; turn halfway during cooking. Remove from heat and keep warm.
3 Brush bread liberally with oil on both sides; toast until golden. Rub cut surface of garlic over bread. Place cooked tomatoes onto bread, top with artichokes and sprinkle with parsley. Serve immediately.

C O O K ' S   F I L E

Storage time: Cook this dish just before serving.
Note: Bruschetta are toasted Italian bread slices flavoured with olive oil and garlic. Any kind of crusty bread can be prepared in this way.

CORN AND CHEESE CHOWDER

Preparation time: 15 minutes
Total cooking time: 30 minutes
Serves 8

90 g (3 1/4 oz) butter
2 large onions, finely chopped
1 garlic clove, crushed
2 teaspoons cumin seeds
1 litre (4 cups) chicken stock
2 medium potatoes, peeled and
    chopped
250 g (1 cup) canned creamed
    corn

400 g (2 cups) fresh corn
    kernels
15 g (1/4 cup) chopped fresh
    parsley
125 g (1 cup) grated Cheddar
    cheese
60 ml (1/4 cup) cream, optional
2 tablespoons chopped fresh
    chives, to garnish

➤ HEAT BUTTER in heavy-based pan. Add onion, cook over medium-high heat 5 minutes, or until golden.
1 Add garlic and cumin seeds and cook 1 minute, stirring constantly. Add chicken stock. Bring to the boil. Add the potato, reduce heat. Simmer,

uncovered, 10 minutes.
2 Add creamed corn, corn kernels and parsley. Bring to boil, reduce heat and simmer for 10 minutes more.
3 Stir through cheese and cream and season to taste. Heat gently until cheese melts. Serve immediately, sprinkled with chopped chives.

C O O K ' S   F I L E

Storage time: Cook this dish up to one day in advance. Reheat and add cheese just before serving.
Variation: Corn kernels scraped from fresh young corn on the cob are best for this recipe—frozen or canned corn may be used if fresh is unavailable.

Grilled Tomatoes with Bruschetta (top) and Corn and Cheese Chowder

EGGPLANT AND ZUCCHINI POTS WITH CAPSICUM RELISH

Preparation time: 12 minutes +
  20 minutes standing
Total cooking time: 40 minutes
Makes 6

1 large eggplant (aubergine),
  cut into 1 cm (1/2 inch) cubes
200 g (7 oz) fresh ricotta cheese
310 g (1 1/4 cups) sour cream
3 eggs
1 tablespoon cornflour
  (cornstarch)
135 g (1 cup) grated zucchini
  (courgette)
1/2 teaspoon cracked black pepper

Capsicum Relish
185 ml (3/4 cup) brown vinegar
90 g (1/3 cup) sugar
1 teaspoon yellow mustard seeds
1 green apple, chopped
1 pear, chopped
1 red capsicum (pepper), chopped
1 green capsicum (pepper),
  chopped

➤ PREHEAT OVEN to 210°C (415°F/ Gas 6–7). Brush six 60 ml (1/4-cup) capacity ramekins with oil. Place eggplant in colander and sprinkle with 1 tablespoon salt; leave 20 minutes. Rinse under cold water; drain well.

1 Using electric beaters, beat ricotta and cream until light and creamy. Add eggs and cornflour, beat until smooth. Transfer to large mixing bowl and gently fold in eggplant, zucchini and black pepper.

2 Spoon mixture evenly into pots. Arrange in a deep baking dish. Fill dish two-thirds up side of pots with warm water; cover loosely with foil. Bake 40 minutes, or until a skewer comes out clean when inserted in centre. Serve with Capsicum Relish.

3 To make Capsicum Relish: Heat vinegar, sugar and mustard seeds for 5 minutes, or until sugar dissolves and mixture boils. Add remaining ingredients. Bring to boil, reduce heat and simmer, uncovered, 30 minutes.

COOK'S FILE

Storage time: Cook this dish up to one hour before serving. Relish can be made up to two days ahead.

1

2

3

TWO-CHEESE RISOTTO CAKES

Preparation time: 30 minutes +
   1 hour 15 minutes refrigeration
Total cooking time: 30 minutes
Serves 6

810 ml (3¼ cups) chicken stock
1 tablespoon olive oil
20 g (½ oz) butter
1 small onion, finely chopped
275 g (1¼ cups) short-grain rice
35 g (⅓ cup) freshly grated
   Parmesan cheese
30 g (1 oz) mozzarella cheese,
   cut into 1 cm (½ inch) cubes
35 g (1¼ oz) sun-dried (sun-
   blushed) tomatoes, chopped
oil, for deep-frying
70 g (2½ oz) mixed salad
   leaves, to serve

➤ BOIL STOCK in small pan. Reduce heat, cover; keep gently simmering.
1 Heat oil and butter in a medium heavy-based pan. Add onion, stir over medium heat 3 minutes until golden; add rice. Reduce heat to low, stir 3 minutes, or until lightly golden. Add a quarter of the stock to pan. Stir 5 minutes, or until all liquid is absorbed.
2 Repeat process until all stock has been added and rice is almost tender, stirring constantly. Stir in Parmesan. Remove from heat. Transfer to a bowl to cool; refrigerate 1 hour.
3 With wetted hands, roll 2 table-spoonfuls of rice mixture into a ball. Make an indentation into the ball, and press in a cube of mozzarella and a couple of pieces of sun-dried tomato. Reshape ball to cover completely, then flatten slightly to a disc shape. Refrigerate for 15 minutes.
4 Heat oil in a medium heavy-based pan. Gently lower risotto cakes a few at a time into moderately hot oil. Cook 1–2 minutes, or until golden brown. Remove with a slotted spoon; drain on paper towels. To serve, arrange salad leaves on each plate and place three risotto cakes on top. Serve immediately.

COOK'S FILE

Storage time: Two-Cheese Risotto Cakes can be prepared up to one day in advance. Store in refrigerator. Fry just before serving.

1

2

3

4

CHILLI PUFFS WITH CURRIED VEGETABLES

Preparation time: 20 minutes
Total cooking time: 1 hour 5 minutes
Makes 12

90 g (3¼ oz) butter
155 g (1¼ cups) plain
    (all-purpose) flour, sifted
¼ teaspoon chilli powder
4 eggs, lightly beaten
4 yellow squash
100 g (3½ oz) snow peas
    (mangetout)
1 carrot
50 g (1¾ oz) butter, extra
2 medium onions, sliced
2 tablespoons mild curry paste
300 g (10½ oz) small oyster
    mushrooms
1 tablespoon lemon juice

➤ PREHEAT OVEN to 240°C (475°F/ Gas 8). Grease and line two 32 x 28 cm (13 x 11 inch) oven trays. Combine the butter and 310 ml (1¼ cups) water in a medium pan. Stir over low heat 5 minutes, or until butter has melted and mixture reaches the boil. Remove from heat; add flour and chilli all at once, stir with a wooden spoon until just combined.

1 Return pan to heat, beating constantly over low heat 3 minutes, or until it thickens and comes away from side and base of pan. Transfer mixture to a large bowl. Using electric beaters, beat mixture on high speed for 1 minute. Add eggs gradually, beating until mixture is glossy. (This stage could take up to 5 minutes.)

2 Place 2 heaped tablespoons of mixture at a time onto prepared trays about 10 cm (4 inches) apart. Bake on top shelf for 20 minutes. Reduce heat to 210°C (415°F/Gas 6–7) and bake for

30 minutes, or until crisp and well browned. (Cut a small slit into each puff halfway during cooking to allow excess steam to escape and puff to dry out.) Cool puffs on a wire rack.

3 Slice squash thinly. Cut snow peas in half diagonally. Cut carrot into thin strips. Heat butter in pan, add onion. Cook over low heat 5 minutes, or until golden; stir in curry paste. Add mushrooms and vegetables, stir over high heat for 1 minute. Add lemon juice, remove from heat, stir.

Cut puffs in half (see Note) and fill with vegetables. Serve immediately.

COOK'S FILE

Storage time: Cook this dish just before serving. Unfilled puffs can be frozen for up to three months.

Note: A small amount of uncooked choux mixture in the centre of cooked curry puffs is a result of the large size of the puffs. Remove mixture using a spoon, then dry out the shells in a warm oven.

CREAMED FENNEL SOUP

Preparation time: 10 minutes
Total cooking time: 35 minutes
Serves 4

2 potatoes
1 medium fennel bulb
60 g (2¼ oz) butter
500 ml (2 cups) chicken stock
125 g (4½ oz) cream cheese,
   chopped
1 tablespoon chopped fresh
   chives
1 tablespoon lemon juice

➤ CHOP potatoes.
1 Slice and chop fennel. Heat butter in a medium pan; add fennel. Cook, covered, over low heat for 10 minutes, stirring occasionally. Do not allow fennel to colour. Add the potatoes and stock to pan, stir. Bring to the boil, reduce heat to low. Cover and cook for 10 minutes, or until veg-etables are tender. Season to taste with salt and ground black pepper. Remove from heat; cool slightly.
2 Transfer mixture to a food processor bowl; add cheese. Process until mixture is smooth and creamy.
3 Return soup to pan. Add chives and juice, stir over low heat until just heated through.

COOK'S FILE

Storage time: Soup can be made one day ahead. Store in refrigerator.

GREEN PEA SOUP

Preparation time: 20 minutes +
2 hours soaking
Total cooking time: 1 hour 40 minutes
Serves 4–6

330 g (1¹/2 cups) green split
peas
2 tablespoons oil
1 onion, finely chopped
1 stalk celery, finely sliced
1 carrot, finely sliced
1 tablespoon ground cumin
1 tablespoon ground coriander
2 teaspoons grated fresh ginger

1.25 litres (5 cups) chicken
stock
310 g (2 cups) frozen green peas
1 tablespoon chopped fresh mint
90 g (¹/3 cup) plain yoghurt or
sour cream, to serve

➤ SOAK SPLIT peas in cold water
for 2 hours.
1 Drain peas well. Heat oil in a large
heavy-based pan, add onion, celery
and carrot. Cook over medium heat for
3 minutes, stirring occasionally, until
soft but not browned. Stir in cumin,
coriander and ginger, cook 1 minute.
2 Add split peas and chicken stock to
pan. Bring to boil; reduce heat to low.

Simmer, covered, for 1¹/2 hours, stirring occasionally.
3 Add frozen peas to pan and stir to
combine; set aside to cool. When cool,
purée soup in batches in a blender or
food processor until smooth. Return to
pan, gently reheat. Season to taste.
Stir in mint. Serve with a swirl of
yoghurt or sour cream in each bowl.

COOK'S FILE

Storage time: Soup may be made up
to one day in advance. Store in refrigerator. Reheat gently and stir in mint
just before serving.
Hint: If soup becomes too thick after
refrigeration, thin with extra stock.

RICH RED ONION SOUP

Preparation time: 20 minutes
Total cooking time: 50 minutes
Serves 6

1 tablespoon oil
20 g (¹/2 oz) butter
1 kg (2 lb 4 oz) red onions
1 tablespoon plain (all-purpose)
flour
1 litre (4 cups) beef stock
250 ml (1 cup) red wine
250 ml (1 cup) puréed
tomato

French bread, cut diagonally
into 12 slices, 2 cm- (³/4 inch)
thick
60 g (¹/2 cup) finely grated
Cheddar cheese

➤ HEAT OIL and butter in a large
heavy-based pan.
1 Slice onions thinly. Add onions, stir-
fry over a high heat for 3 minutes
until soft and starting to become gold-
en. Reduce heat to medium-low, and
cook onions 25 minutes, stirring occasionally, until very soft and golden.
2 Sprinkle flour over onions, stir well
with a wooden spoon. Cook for 2 minutes, stirring constantly. Add stock,

wine and puréed tomato, stir until
mixture boils and thickens slightly.
Season. Reduce heat to low and simmer, uncovered, for 20 minutes.
3 Toast bread on both sides under a
griller (broiler). Top with grated
cheese, return to grill and cook until
melted and golden. To serve, ladle
soup into deep bowls and float bread
slices on top.

COOK'S FILE

Storage time: Rich Red Onion Soup
may be made up to one day ahead.
Store in refrigerator. Reheat soup gently, toast bread and grill cheese just
before serving.

Green Pea Soup (top)
and Rich Red Onion Soup.

MUSHROOM CAPS WITH GARLIC AND THYME

Preparation time: 20 minutes
Total cooking time: 25 minutes
Serves 6

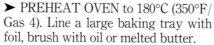

6 large flat or field mushrooms
    (about 80 g/2³/4 oz each)
2 tablespoons oil
1 small onion, finely chopped
3 rashers bacon, finely chopped
2 garlic cloves, crushed
4 slices white bread
1 tablespoon fresh thyme leaves

➤ PREHEAT OVEN to 180°C (350°F/Gas 4). Line a large baking tray with foil, brush with oil or melted butter.

1 Using your fingers, peel the skin from the mushrooms. Remove the stems and chop finely. Heat oil in a heavy-based frying pan. Add the onion and bacon and cook over medium heat until golden. Add chopped mushroom stems and garlic, cook for 3 minutes over medium heat until soft, stirring occasionally.

2 Transfer to a medium mixing bowl to cool. Remove crusts from bread, tear into pieces and place in food processor bowl. Using pulse action,

process 20 seconds, or until fluffy crumbs form.

3 Add the breadcrumbs and the thyme to the bowl and stir until well combined. Place the mushrooms on the prepared tray and top with the breadcrumb mixture. Grind pepper over. Bake for 20 minutes, or until the mushrooms are tender and breadcrumbs are golden. Serve at once.

COOK'S FILE

Storage time: Topping may be prepared up to four hours in advance. Cook mushrooms just before serving.

1

2

3

INDIVIDUAL SPINACH SOUFFLES

Preparation time: 20 minutes
Total cooking time: 25 minutes
Serves 4

35 g (1/3 cup) dry breadcrumbs
100 g (3 1/2 oz) silverbeet (Swiss chard) leaves
40 g (1 1/2 oz) butter
2 tablespoons plain (all-purpose) flour
250 ml (1 cup) milk
4 eggs, separated
60 g (1/2 cup) finely grated Cheddar cheese
2 tablespoons freshly grated Parmesan cheese
pinch cayenne pepper, to taste
1 tablespoon freshly grated Parmesan cheese, extra

➤ PREHEAT OVEN to 180°C (350°F/ Gas 4). Brush four 185 ml (3/4-cup) soufflé dishes with melted butter or oil. Coat base and sides evenly with breadcrumbs, shake off excess.

1 Wash silverbeet leaves thoroughly. Shred finely, and steam for 1 minute, until just tender; cool. Using your hands, squeeze all excess moisture. Spread to separate strands. Set aside.

2 Heat butter in a medium pan; add flour. Stir over a low heat 2 minutes, or until lightly golden. Add milk gradually to pan, stirring until mixture is smooth. Stir constantly over medium heat 2 minutes until mixture boils and thickens; boil 1 minute more. Remove from heat. Add egg yolks, beat until smooth. Add cheeses and cayenne; season with salt. Stir until melted and almost smooth; stir in spinach.

3 Using electric beaters, beat egg whites in a clean dry bowl until stiff peaks form. Using a metal spoon, fold gently into silverbeet mixture.

4 Spoon into prepared soufflé dishes, bake for 20 minutes, or until well risen and browned. Sprinkle with the extra Parmesan cheese and serve immediately.

COOK'S FILE

Storage time: Cook soufflés just before serving. Silverbeet base can be prepared up to one day ahead; fold whites into mixture just before cooking.

1

2

3

4

VEGETABLE TEMPURA

Preparation time: 30 minutes +
  1 hour refrigeration
Total cooking time: 5 minutes per batch
Serves 6

155 g (1¼ cups) plain
  (all-purpose) flour
1 egg
125 g (4½ oz) broccoli
1 small onion
1 small red capsicum (pepper)
1 small green capsicum (pepper)
1 carrot
50 g (1¾ oz) green beans
light vegetable oil, for deep-
  frying

Dipping Sauce
80 ml (⅓ cup) soy sauce
2 tablespoons Thai sweet chilli
  sauce
1 garlic clove, crushed
1 tablespoon honey

➤ SIFT FLOUR into mixing bowl.
1 Make a well in the centre, add egg
and 310 ml (1¼ cups) water and
whisk until combined. Cover and
refrigerate 1 hour.
2 Cut broccoli into small florets.
Finely slice onion, and cut capsicum
and carrot into thin strips about 6 cm
(2½ inches) long. Cut beans to about
6 cm (2½ inches) long, and halve
lengthways. Add vegetables to batter
and stir to combine.

3 Heat oil in a pan. Using tongs, gather a small bunch of batter-coated vegetables (roughly two pieces of each vegetable) and lower into oil. Hold submerged for a few seconds until batter begins to set and vegetables hold together. Release from tongs and cook until crisp and golden. Drain on paper towels. Repeat until all vegetables are cooked. Serve immediately with dipping sauce.
To make Dipping Sauce: Gently whisk together all the ingredients in a small bowl with a fork until combined.

COOK'S FILE

Storage time: Vegetables may be prepared up to four hours in advance. Batter and cook just before serving.

GRILLED EGGPLANT AND ASPARAGUS SANDWICH

Preparation time: 45 minutes +
   30 minutes standing
Total cooking time: 40 minutes
Serves 6

2 eggplants (aubergines)
1 tablespoon olive oil
1 bunch asparagus
50 g (1³/4 oz) butter
40 g (¹/3 cup) finely chopped
   spring onions (scallions)
60 g (¹/2 cup) plain (all-purpose)
   flour
250 ml (1 cup) milk
30 g (¹/4 cup) grated Romano
   cheese
100 g (3¹/2 oz) prosciutto, finely
   shredded
1 tablespoon lemon juice
1 egg yolk

Soufflé Topping
1 egg white
30 g (¹/4 cup) grated Romano
   cheese

➤ PREHEAT OVEN to 180°C (350°F/ Gas 4). Line an oven tray with foil. Brush with melted butter or oil. Cut eggplant into six slices lengthways. Sprinkle with salt, stand 30 minutes. Rinse under cold water, drain. Pat dry with paper towels.

1 Place eggplant on cold grill (broiler) tray. Brush with oil. Cook under medium-high heat 5 minutes, or until golden brown each side. Drain on paper towels. Trim woody ends from asparagus. Steam or microwave until just tender. Cut six asparagus into 5 cm (2 inch) lengths, set aside. Finely chop remaining asparagus. Heat butter in pan. Add onions, cook over medium-high heat 1 minute. Stir in flour. Add milk gradually to pan, stirring until mixture is smooth. Stir constantly over medium heat 5 minutes, or until it boils and thickens; boil for 1 minute.

2 Add cheese, prosciutto, juice, yolk and chopped asparagus. Season with pepper. Mix well. Remove from heat. Place six eggplant slices on prepared tray. Spread asparagus filling evenly over each eggplant slice. Top with remaining eggplant slices.

3 To make Soufflé Topping: Using electric beaters, beat egg white in small, clean bowl until stiff peaks form. Spread egg white evenly over eggplant. Sprinkle with extra cheese, decorate with remaining asparagus. Bake for 15 minutes until cheese is melted. Serve immediately.

COOK'S FILE

Storage time: This recipe can be assembled up to eight hours ahead of time. Prepare Soufflé Topping just before cooking.

Hint: Romano cheese and prosciutto (also known as Parma ham) are available from delicatessens.

1

2

3

RED CAPSICUM SOUP

Preparation time: 20 minutes
Total cooking time: 50 minutes
Serves 6

4 red capsicums (peppers)
4 tomatoes
60 ml (¼ cup) oil
½ teaspoon dried marjoram
½ teaspoon dried mixed
    herbs
2 garlic cloves, crushed
1 teaspoon mild curry paste
1 red onion, sliced
1 medium leek, sliced
    (white part only)

250 g (9 oz) green cabbage,
    chopped
1 teaspoon sweet chilli sauce

➤ CUT CAPSICUM in quarters.
Remove seeds and membrane.
1 Place skin-side up on grill (broiler)
tray. Brush with a little oil. Grill (broil)
for 10 minutes, or until skin is black.
Cover with a damp tea towel until cool.
Peel off skin. Mark a small cross on the
top of each tomato. Place in a bowl of
boiling water for 1–2 minutes, then
plunge into cold water. Peel skin off
downwards from the cross. Cut tomatoes in half; gently squeeze out seeds.
2 Heat oil in a large pan. Add herbs,
garlic and curry paste. Stir over low

heat for 1 minute, or until aromatic.
Add onion and leek and cook 3 minutes, or until light golden. Add
cabbage, tomatoes, capsicum and
1 litre (4 cups) water. Bring to boil,
reduce heat and simmer 20 minutes.
Remove from heat. Cool slightly.
3 Place soup in small batches in a
food processor bowl. Using pulse
action, process 30 seconds, or until
smooth. Return soup to clean pan, stir
through chilli sauce, season to taste.
Reheat gently. Serve hot.

COOK'S FILE

Storage time: This soup can be
made up to five days ahead and
refrigerated. Reheat before serving.

ARTICHOKES WITH TARRAGON MAYONNAISE

Preparation time: 30 minutes
Total cooking time: 30 minutes
Serves 4

4 medium globe artichokes
60 ml (1/4 cup) lemon juice

Tarragon Mayonnaise
1 egg yolk
1 tablespoon tarragon vinegar
1/2 teaspoon French mustard
170 ml (2/3 cup) olive oil

➤ TRIM STALKS from the base of the artichokes.

1 Using scissors, trim points from outer leaves. Using a sharp knife, cut top from artichoke. Brush all cut areas of artichokes with lemon juice to prevent discolouration.

2 Steam artichokes for 30 minutes, until tender. Top up pan with boiling water if necessary. Remove from heat and set aside to cool.

3 To make Tarragon Mayonnaise: Place egg yolk, vinegar and mustard in a medium mixing bowl. Using a wire whisk, beat for 1 minute. At first, add oil a teaspoon at a time, whisking constantly until mixture is thick and creamy. As the mayonnaise thickens, pour oil in a thin, steady stream. Continue whisking until all the oil is added. Season to taste. Place a cooled artichoke on each plate with a little Tarragon Mayonnaise.

COOK'S FILE

Storage time: Artichokes may be cooked up to four hours in advance. Mayonnaise may be made up to two hours in advance and refrigerated.

Hint: To eat artichokes, take off a leaf at a time, dip base of leaf in mayonnaise and scrape off fleshy base with teeth. Towards the centre of the artichoke, the leaves are more tender and more of the leaf is edible. Provide a bowl for discarded leaves.

MAIN COURSES

MEXICAN-STYLE VEGETABLES

Preparation time: 30 minutes +
  2 hours refrigeration
Total cooking time: 50 minutes
Serves 6

Polenta
330 ml (1¹/₃ cups) chicken stock
150 g (1 cup) polenta
50 g (¹/₂ cup) freshly grated
  Parmesan cheese
2 tablespoons olive oil

1 large green capsicum (pepper)
1 large red capsicum (pepper)
3 tomatoes
6 green button squash
1 fresh corn cob
1 tablespoon oil
1 onion, sliced
1 tablespoon ground cumin
¹/₂ teaspoon chilli powder
2 tablespoons chopped fresh
  coriander (cilantro)

➤ BRUSH a 20 cm (8 inch) round springform tin with oil.
1 To make Polenta: Place chicken stock and 250 ml (1 cup) water in a medium pan and bring to the boil.

Add the polenta and stir to combine; stir constantly for 10 minutes until very thick (see Note). Remove from heat and stir in Parmesan. Spread the mixture into the prepared tin and smooth the surface. Refrigerate for 2 hours. Turn out, then cut into six wedges. Brush one side with oil, cook under preheated grill (broiler) for 5 minutes, or until edges are browned. Repeat with other side.
2 Cut the capsicum into 2 cm (³/₄ inch) squares, chop tomatoes, cut squash into quarters and cut corn into 2 cm (³/₄ inch) slices, quartered.
3 Heat oil in large pan. Cook onion over medium heat 5 minutes, until soft. Stir in the cumin and chilli powder; cook 1 minute. Add vegetables. Bring to the boil, reduce heat. Simmer, covered, over low heat 30 minutes, or until vegetables are tender, stirring occasionally. Stir in the coriander and season with salt and pepper. Serve hot with wedges of polenta.

COOK'S FILE

Storage time: Vegetables can be cooked up to one day ahead. Polenta can be cooked one day ahead. Grill just before serving.
Note: Polenta must be stirred for the time given, otherwise it will be gritty.

FETTUCCINE WITH CREAMY MUSHROOM AND BEAN SAUCE

Preparation time: 15 minutes
Total cooking time: 20 minutes
Serves 4

100 g (2/3 cup) pine nuts
375 g (13 oz) fettuccine
250 g (9 oz) green beans
2 tablespoons oil
1 onion, chopped
2 garlic cloves, crushed
250 g (9 oz) mushrooms, sliced
125 ml (1/2 cup) white wine
310 ml (1 1/4 cups) cream
125 ml (1/2 cup) vegetable stock

1 egg
15 g (1/4 cup) chopped fresh basil
35 g (1/4 cup) sun-dried (sun-blushed) tomatoes, cut into thin strips
50 g (1 3/4 oz) Parmesan cheese, shaved

➤ PLACE PINE NUTS in small pan. Stir over medium heat until golden. Set aside. Cook fettuccine in a large pot of boiling water with a little oil added until just tender. Drain and keep warm.

1 Trim tops and tails of beans and cut into long thin strips.

2 Heat oil in a large heavy-based frying pan. Cook onion and garlic over medium heat 3 minutes, or until softened. Add mushrooms, cook, stirring, for 1 minute. Add wine, cream and stock. Bring to boil, reduce heat, simmer for 10 minutes.

3 Lightly beat egg in a bowl. Stirring constantly, add a little cooking liquid to the egg. When combined, pour mixture slowly into pan, stirring constantly for 30 seconds. Add beans, basil, pine nuts and sun-dried tomatoes; stir until heated through. Season. Divide pasta among warmed serving plates and pour sauce over. Garnish with Parmesan. Serve immediately.

COOK'S FILE

Storage time: Cook this dish just before serving.

VEGETABLE LASAGNE

Preparation time: 20 minutes
Total cooking time: 1 hour 15 minutes
Serves 6

3 large red capsicum (peppers)
2 large eggplants (aubergines)
2 tablespoons oil
1 large onion, finely chopped
3 garlic cloves, crushed
1 teaspoon dried mixed
   herbs
1 teaspoon dried oregano
500 g (1 lb 2 oz) mushrooms,
   sliced
440 g (15¹/₂ oz) can whole
   tomatoes, crushed
440 g (15¹/₂ oz) can red kidney
   beans, drained
1 tablespoon sweet chilli sauce
250 g (9 oz) packet instant
   lasagne sheets
1 bunch English spinach,
   chopped
50 g (1 cup) fresh basil leaves
100 g (3¹/₂ oz) sun-dried (sun-
   blushed) tomatoes, sliced
25 g (¹/₄ cup) grated Parmesan
   cheese
30 g (¹/₄ cup) grated Cheddar
   cheese

Cheese Sauce
60 g (2¹/₄ oz) butter
30 g (¹/₄ cup) plain (all-purpose)
   flour
500 ml (2 cups) milk
600 g (1 lb 5 oz) ricotta cheese

➤ PREHEAT OVEN to 180°C (350°F/
Gas 4). Grease a 35 x 28 cm (14 x
11 inch) 8–10-cup capacity ovenproof
casserole dish.

1 Cut capsicum in quarters. Remove
seeds and membrane. Place skin-side
up on grill (broiler) tray. Brush with
oil. Grill (broil) 10 minutes, or until
skin is black. Cover with damp tea
towel until cool. Peel off skin. Cut capsicum into long thin strips. Set aside.
Slice eggplant into 1 cm (¹/₂ inch)
rounds. Cook in pan of boiling water
1 minute, or until just tender, drain.
Pat dry with paper towels. Set aside.

2 Heat oil in frying pan. Cook onion,
garlic and herbs over medium heat for
5 minutes, or until onion is soft. Add
mushrooms, cook 1 minute. Add
tomatoes, beans and chilli sauce.
Season. Bring to boil, reduce heat.
Simmer, uncovered, 15 minutes, or until
thick. Remove from heat. Dip lasagne
sheets in hot water to soften slightly
and arrange four sheets in the dish.

3 Arrange half of each of the eggplant,

spinach, basil, capsicum, mushroom
mixture and sun-dried tomatoes over
pasta. Top with pasta; press gently.
Repeat layers. Top with cheese sauce,
sprinkle with combined cheeses. Bake
for 45 minutes, or until pasta is soft.

4 To make Cheese Sauce: Heat
butter in pan, add flour. Stir over medium heat for 2 minutes, or until golden.
Add milk gradually, stirring until mixture boils and thickens. Boil 1 minute.
Add ricotta, stir until smooth.

VEGETABLE STIR-FRY

Preparation time: 15 minutes
Total cooking time: 5 minutes
Serves 4

2 spring onions (scallions)
250 g (9 oz) broccoli
1 red capsicum (pepper)
1 yellow capsicum (pepper)
150 g (5½ oz) button mushrooms
1 tablespoon oil
1 teaspoon sesame oil
1 garlic clove, crushed

2 teaspoons grated fresh ginger
35 g (¼ cup) halved pitted black
  olives
1 tablespoon soy sauce
1 tablespoon honey
1 tablespoon sweet chilli sauce
1 tablespoon sesame seeds

➤ FINELY SLICE the spring onions.
Cut the broccoli in small florets.
1 Cut capsicum in halves, remove
seeds and membrane. Cut into thin
strips. Cut mushrooms in half.
2 Heat oils in a wok or large frying
pan. Add garlic, ginger and onions.

Stir-fry over medium heat 1 minute.
Add broccoli, capsicum, mushrooms
and olives. Stir-fry for 2 minutes, or
until vegetables are bright in colour
and just tender.
3 Combine soy sauce, honey and chilli
sauce. Place sesame seeds on an oven
tray, toast under hot grill (broiler) until
golden. Pour sauce over vegetables,
toss lightly to combine. Sprinkle with
sesame seeds and serve immediately.

COOK'S FILE

Storage time: Cook this dish just
before serving.

PUMPKIN GNOCCHI WITH SAGE BUTTER

Preparation time: 30 minutes +
  5 minutes standing
Total cooking time: 1 hour 45 minutes
Serves 4

500 g (1 lb 2 oz) pumpkin
185 g (1½ cups) plain
  (all-purpose) flour
25 g (¼ cup) freshly grated
  Parmesan cheese

Sage Butter
100 g (3½ oz) butter
2 tablespoons chopped fresh
  sage

25 g (¼ cup) freshly grated
  Parmesan cheese, extra

➤ PREHEAT OVEN to 180°C (350°F/
Gas 4). Brush a baking tray with oil or
melted butter.
1 Cut pumpkin into large pieces and
place on prepared tray. Bake for
1½ hours, until very tender. Cool
slightly. Scrape flesh from skin, avoid-
ing any tough or crispy parts. Place
into a large mixing bowl. Sift flour
into bowl, add Parmesan cheese and
pepper. Mix until well combined. Turn
onto a lightly floured surface, knead
for 2 minutes, or until smooth.
2 Divide dough in half. Using floured
hands, roll each half into a sausage
40 cm (16 inches) long. Cut into 16 equal

pieces. Form each piece into an oval
shape, indent with floured fork prongs.
3 Heat a large pan of water until boiling. Gently lower batches of gnocchi
into water. Cook until gnocchi rise to
the surface, and then 3 minutes more.
Drain and keep warm.
To make Sage Butter: Melt butter
in small pan, remove from heat and
stir in sage. Set aside for 5 minutes to
keep warm.
To serve, divide gnocchi among
bowls, drizzle with Sage Butter and
sprinkle with extra Parmesan cheese.

COOK'S FILE

Storage time: Gnocchi can be prepared up to four hours ahead. Cook
just before serving.

Vegetable Stir-fry (top)
and Pumpkin Gnocchi with Sage Butter.

POTATO CAKES WITH PRAWNS AND ZUCCHINI

Preparation time: 20 minutes
Total cooking time: 30 minutes
Serves 6

3 zucchini (courgettes)
310 g (2 cups) grated potato
1 egg, lightly beaten
1 teaspoon ground paprika
60 g (1/2 cup) plain (all-purpose) flour
60 ml (1/4 cup) oil
60 g (21/4 oz) butter
1/2 bunch chives, chopped
1 tablespoon chopped fresh dill

60 g (1/2 cup) plain (all-purpose) flour, extra
500 ml (2 cups) milk
15 g (1/4 cup) chopped parsley
1 tablespoon lime juice
500 g (1 lb 2 oz) school prawns (shrimp), peeled

➤ CUT ZUCCHINI into 5 mm (1/4 inch) slices. Drain potato; pat dry with paper towels.

1 Combine potato, egg and paprika. Shape 3 tablespoons mixture into balls. Roll in flour and flatten slightly. Heat oil in frying pan. Fry in batches over medium heat 2–3 minutes each side, or until golden. Drain on paper towels.

2 Heat butter in medium pan. Cook chives and zucchini over medium heat 5 minutes, or until tender. Remove, drain. Add dill and extra flour to pan. Stir over low heat 2 minutes, or until flour is lightly golden.

3 Add milk gradually to pan, stirring until smooth. Stir constantly over medium heat 4 minutes, or until it boils and thickens. Boil for 1 minute. Add parsley and lime juice. Season. Stir in prawns, chives and zucchini. Heat gently. Pour over potato cakes.

COOK'S FILE

Storage time: Prawn/zucchini mixture can be prepared a day ahead. Store in refrigerator, reheat gently.
Variation: Omit prawns or replace with another sliced green vegetable. Cook with the chives and zucchini.

CARROT PESTO SLICE

Preparation time: 45 minutes
Total cooking time: 50 minutes +
  30 minutes standing
Serves 4

50 g (1³/4 oz) butter
60 g (¹/2 cup) plain (all-purpose)
  flour
750 ml (3 cups) milk
160 g (²/3 cup) light sour cream
1 teaspoon cracked black
  pepper
100 g (3¹/2 oz) Cheddar cheese,
  grated
4 eggs, lightly beaten
2 tablespoons bottled pesto

750 g (1 lb 10 oz) carrots,
  peeled and grated
250 g (9 oz) packet instant
  lasagne sheets
50 g (1³/4 oz) Cheddar cheese,
  grated, extra

➤ GREASE a 30 x 20 cm (12 x 8 inch)
ovenproof baking dish.

1 Heat butter in large pan; add flour.
Stir over low heat until mixture is
lightly golden and bubbling. Add
combined milk, sour cream and pepper gradually to pan, stirring until
mixture is smooth. Stir constantly
over medium heat 5 minutes, or until
it boils and thickens; boil 1 minute,
remove from heat. Stir in cheese, cool
slightly. Add beaten eggs gradually,

stirring constantly.

2 Pour one-third of the sauce into
another bowl for topping; set aside.
Add pesto and carrot to remaining
sauce, stirring to combine.

3 Preheat oven to 150°C (300°F/Gas 2).
Beginning with a layer of the carrot
mixture, alternate layers of carrot with
lasagne sheets in prepared dish. Use
three layers of each, finishing with
pasta. Spread reserved sauce evenly
over the top. Sprinkle with extra
cheese. Leave for 15 minutes before
cooking (to allow pasta to soften).
Bake for 40 minutes, or until set and
firm to touch. Remove from the oven,
cover and set aside for 15 minutes
prior to serving (this will help it to
slice cleanly).

CAPSICUM FRITTATA

Preparation time: 15 minutes
Total cooking time: 40 minutes
Serves 4

660 g (1 lb 7 oz) jar pimiento
  pieces
4 rashers bacon, optional
1 tablespoon olive oil
2 medium red onions, finely
  chopped
6 eggs, lightly beaten
60 g (1/2 cup) grated Cheddar
  cheese

50 g (1/2 cup) grated Parmesan
  cheese
1 tablespoon plain (all-purpose)
  flour
15 g (1/4 cup) chopped fresh
  parsley

➤ PREHEAT OVEN to 180°C (350°F/
Gas 4). Grease a 23 cm (9 inch) ovenproof pie plate.
1 Rinse pimiento pieces, drain. Pat
with paper towels. Cut into thin strips.
Trim fat from bacon; place bacon on a
cold grill (broiler) tray. Cook under
medium-high heat until crisp. Drain
on paper towels. Cut into small pieces.

2 Heat oil in medium pan. Add onions,
cook over medium heat for 2 minutes.
Remove; drain on paper towels.
3 Combine eggs, pimiento, bacon and
onions. Add combined cheeses, flour
and parsley. Season with salt and
1/2 teaspoon ground black pepper. Mix
well. Spoon into prepared pie plate.
Bake for 25–30 minutes, or until set
and firm to touch. Serve hot or cold
with a green salad and crusty bread.

COOK'S FILE

Hint: Pimiento is bottled red capsicum (pepper). Substitute 2 fresh red
capsicum, grilled (broiled), if desired.

1

2

3

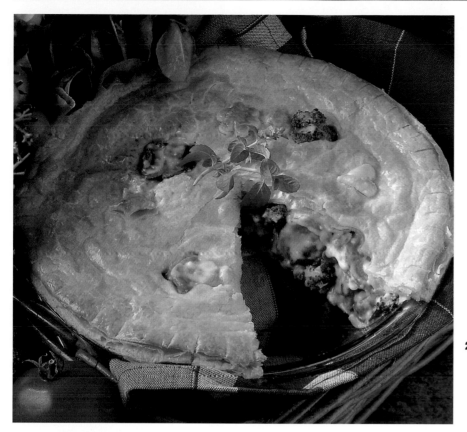

VEGETABLE PIE

Preparation time: 40 minutes +
    30 minutes standing
Total cooking time: 50 minutes
Serves 6

1 small eggplant (aubergine)
30 g (1 oz) butter
2 garlic cloves, crushed
2 spring onions (scallions), sliced
200 g (7 oz) orange sweet
    potato, cut into 1 cm
    (1/2 inch) cubes
1 carrot, thinly sliced
50 g (1³/4 oz) button
    mushrooms, sliced
1 small red capsicum (pepper),
    finely sliced
150 g (5¹/2 oz) broccoli, cut in
    small florets
40 g (1¹/2 oz) butter, extra
2 tablespoons plain (all-purpose)
    flour
375 ml (1¹/2 cups) milk
125 g (4¹/2 oz) feta cheese,
    crumbled
25 g (¹/4 cup) grated Parmesan
    cheese
2 tablespoons pine nuts, toasted

1 teaspoon dried oregano
2 eggs, lightly beaten
2 sheets ready-rolled puff
    pastry
1 egg, extra, lightly beaten

➤ BRUSH a 23 cm (9 inch) round pie
dish with melted butter or oil.
1 Cut eggplant into 2 cm (³/4 inch)
cubes. Place in a colander, sprinkle
with salt. Leave for 30 minutes. Rinse;
drain. Pat dry with paper towels.
Preheat oven to 180°C (350°F/Gas 4).
2 Heat butter in large pan. Cook garlic
and spring onion over medium heat
1 minute. Add sweet potato, carrot and
mushrooms, stirring, 4 minutes, or until
just tender. Add capsicum and broccoli. Cook 3 minutes. Add eggplant,
cook 2 minutes. Remove from heat.
3 Melt extra butter in medium pan.
Add flour, stir over medium heat
1 minute, or until mixture is lightly
golden and bubbling. Add milk gradually to pan, stirring until mixture is
smooth. Stir constantly over medium
heat 4 minutes, or until it boils and
thickens. Add cheeses, pine nuts and
oregano. Combine vegetable mixture
and sauce in a bowl. Add eggs, stir
until combined. Spoon into dish.

4 Cut a long strip of pastry slightly
wider than rim of pie dish. Brush rim
of the dish with water and press down
strip. Brush the strip with water.
Place pastry sheet over the top, press
edges to seal. Trim and decorate edge.
Cut out shapes from pastry top to
allow steam to escape and use the cut-
outs to decorate the pie top. Brush top
with beaten egg. Bake for 35–40 minutes, or until the pastry is golden
brown and puffed.

GOURMET VEGETABLE PIZZA

Preparation time: 30 minutes +
  30 minutes standing
Total cooking time: 1 hour 10 minutes
Serves 6

Pizza Dough
20 g (1/3 cup) fresh basil leaves
2 tablespoons polenta or
  cornmeal
1 teaspoon dried yeast
1 teaspoon sugar
185 g (1 1/2 cups) plain
  (all-purpose) flour
125 ml (1/2 cup) warm water
1 teaspoon salt
1 tablespoon olive oil

Tomato Sauce
1 tablespoon oil
1 small red onion, finely
  chopped
1 garlic clove, crushed
1 large tomato, finely chopped
1 tablespoon tomato paste (purée)
1/2 teaspoon dried oregano

Topping
60 g (2 1/4 oz) button mushrooms,
  finely sliced
100 g (3 1/2 oz) fresh baby corn
225 g (1 1/2 cups) grated
  mozzarella cheese
50 g (1 3/4 oz) spinach leaves,
  finely shredded
1 small red capsicum (pepper),
  cut into short thin strips
2 tablespoons pine nuts

➤ FINELY CHOP basil leaves.
1 Brush a 30 cm (12 inch) pizza tray
with oil and sprinkle with polenta.
2 To make Pizza Dough: Combine
yeast, sugar and 2 tablespoons of the
flour in a small mixing bowl.
Gradually add the water; blend until
smooth. Stand, covered, with plastic
wrap, in a warm place for about
10 minutes, or until foamy.
3 Sift the remaining flour into a large
mixing bowl. Add salt and basil, and
make a well in the centre. Add the
yeast mixture and oil. Using a knife,
mix to a soft dough.
4 Turn dough onto a lightly floured
surface, knead for 5 minutes, or until
smooth. Shape dough into a ball, place
in a large, lightly oiled mixing bowl.
Leave, covered with plastic wrap, in a
warm place for 20 minutes, or until
well risen. Meanwhile, prepare sauce.
5 To make Tomato Sauce: Heat
oil in a small pan, add onion and garlic
and cook over a medium heat for
3 minutes, or until soft. Add tomato
and reduce heat; simmer 10 minutes,
stirring occasionally. Stir in tomato
paste and oregano, cook for 3 minutes.
Allow sauce to cool before using.
6 Preheat oven to 210°C (415°F/
Gas 6–7). Turn dough out onto a lightly floured surface and knead a further
5 minutes until smooth. Roll out to fit
the prepared tray. Spread the sauce
onto the pizza base, arrange the mushrooms and corn evenly on top.
Sprinkle over half the cheese, top with
the spinach, capsicum and remaining
cheese. Sprinkle with pine nuts. Bake
for 40 minutes, or until crust is gold-
en. Serve pizza immediately.

COOK'S FILE

Storage time: The Tomato Sauce
may be made up to four hours in
advance. Pizza is best prepared and
cooked just before serving.
Hint: Fresh baby corn can usually be
purchased from greengrocers. If it is
unavailable, use canned baby corn or
canned corn kernels.

1

2

3

4

5

6

SWEET SPICED GOLDEN NUGGETS

Preparation time: 35 minutes
Total cooking time: 1 hour
Serves 4

4 golden nugget pumpkins
1 tablespoon olive oil
210 g (1 cup) brown and wild rice blend
30 g (1 oz) butter
1 onion, finely chopped
300 g (10¹/2 oz) orange sweet potato, cut in 1 cm (¹/2 inch) cubes
65 g (1 cup) chopped spring onions (scallions)

2 teaspoons ground cumin
¹/2 teaspoon ground ginger
1 teaspoon ground coriander
1 teaspoon ground turmeric
1 teaspoon garam masala
2 tablespoons currants, soaked in hot water
40 g (¹/3 cup) grated cheese

➤ PREHEAT OVEN to 180°C (350°F/ Gas 4). Slice the top third off each pumpkin horizontally.

1 Scoop out seeds, leaving a deep cavity. Brush lightly with oil. Stand pumpkins in baking dish. Pour in enough water to come halfway up the sides of the pumpkins. Place lid on each. Bake 20 minutes. Remove from water bath, allow to cool. Cook rice in large pan of boiling water until tender. Drain and cool.

2 Heat butter in frying pan. Add onion and sweet potato. Cover and cook over medium heat 5 minutes. Add spring onion and cook, uncovered, 1 minute. Stir in spices, cook 2 minutes. Remove from heat. Fold through rice and drained currants.

3 Spoon filling into pumpkin cavities. Sprinkle with cheese. Place lid on at an angle, then cover with foil. Bake for 20 minutes. Serve hot with salad.

COOK'S FILE

Note: Brown and wild rice mix is available from supermarkets and delicatessens. If unavailable, white or brown rice can be used.

VEGETABLE TART

Preparation time: 30 minutes +
    20 minutes refrigeration
Total cooking time: 1 hour
Serves 6

155 g (1¼ cups) plain
    (all-purpose) flour
90 g (3¼ oz) butter, chopped
2–3 tablespoons iced water

Vegetable Filling
1 small red capsicum (pepper)
1 small green capsicum (pepper)
200 g (7 oz) pumpkin
1 potato
150 g (5½ oz) broccoli
1 carrot
1 tablespoon oil
1 onion, finely sliced
50 g (1¾ oz) butter
30 g (¼ cup) plain (all-purpose)
    flour
250 ml (1 cup) milk
2 egg yolks
125 g (1 cup) grated Cheddar
    cheese

➤ PREHEAT OVEN to 180°C (350°F/Gas 4). Sift flour into large bowl; add butter. Using fingertips, rub butter into flour for 2 minutes until mixture is fine and crumbly. Add almost all the water, mix to firm dough, adding more water if necessary. Turn onto lightly floured surface, press together until smooth. Roll out and line a deep 23 cm (9 inch) fluted tin. Refrigerate 20 minutes. Cut a sheet of greaseproof paper large enough to cover pastry-lined tin. Spread a layer of dried beans evenly over paper. Bake 10 minutes, remove from oven, discard paper and beans. Return to oven for 10 minutes, or until lightly golden. Cool.

1 To make Vegetable Filling: Cut capsicum, pumpkin and potato into 2 cm (¾ inch) squares. Cut broccoli into florets. Cut carrots into 1.5 cm (⅝ inch) slices.

2 Heat oil in a frying pan, add onion and cook over medium heat 5 minutes, until soft and golden. Add capsicum and cook, stirring, 5 minutes until soft. Transfer to a large mixing bowl to cool. Steam or boil remaining vegetables for 3 minutes, until just tender. Drain well, add to bowl and cool.

3 Heat butter in a small pan; add flour. Stir over a low heat 2 minutes, or until flour mixture is lightly golden. Add milk gradually to pan, stirring until mixture is smooth. Stir constantly over medium heat until mixture boils and thickens; boil 1 minute more, remove from heat. Add yolks, beat until smooth. Stir in half the cheese. Pour the sauce over the cooked vegetables, and stir to thoroughly combine. Pour the mixture into the pastry shell and sprinkle with the remaining cheese. Bake for 25 minutes, or until top is golden.

BRAISED CABBAGE TURNOVERS

Preparation time: 15 minutes
Total cooking time: 1 hour 5 minutes
Makes 6

60 g (2¼ oz) butter
2 onions, thinly sliced
1 stalk celery, thinly sliced
600 g (1 lb 5 oz) green cabbage,
   finely shredded
6 spring onions (scallions),
   finely chopped
15 g (¼ cup) chopped fresh
   parsley
¼ teaspoon ground chilli
   powder
100 g (3½ oz) feta cheese,
   crumbled
6 sheets ready-rolled puff
   pastry
1 egg, lightly beaten

➤ PREHEAT OVEN to 210°C (415°F/
Gas 6–7). Line two 32 x 28 cm (13 x 11
inch) oven trays with baking paper.
1 Heat butter in medium pan; add
onions and celery. Cook over low heat
15 minutes, stirring occasionally. Add
cabbage and 60 ml (¼ cup) water. Stir
over high heat 10 minutes, or until
cabbage has wilted and almost all
liquid has evaporated.
2 Add spring onions, parsley and
chilli to pan; stir. Remove pan from
heat; cool mixture slightly. Add
cheese, mix well; season to taste, cool.
3 Brush each pastry sheet with egg.
Divide cabbage mixture evenly into six
portions. Place one portion of mixture
at a time, slightly off-centre, on pastry
square. Fold sheet in half, press to seal
edges. Cut pastry into a half circle
using an 18 cm (7 inch) saucepan lid as
a guide; discard excess pastry. Repeat
with remaining pastry and filling.
4 Brush tops of half circles with

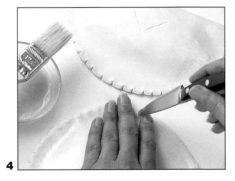

remaining egg. Using a sharp knife,
cut a diamond pattern across top of
pastry. (Do not cut through the pastry.) Use your finger and the back of
knife to decorate the pastry edge.
Arrange the pies on prepared trays.
Bake for 40 minutes, or until puffed
and browned. Serve warm or cool.

COOK'S FILE

Storage time: This dish can be
made up to one day ahead.

SWEET AND SOUR NOODLES AND VEGETABLES

Preparation time: 12 minutes
Total cooking time: 15 minutes
Serves 4–6

200 g (7 oz) thin fresh egg
   noodles
4 fresh baby corn
60 ml (¼ cup) oil
1 green capsicum (pepper),
   sliced
1 red capsicum (pepper), sliced
2 stalks celery, sliced diagonally
1 carrot, sliced diagonally

250 g (9 oz) button mushrooms,
   sliced
3 teaspoons cornflour
   (cornstarch)
2 tablespoons brown vinegar
1 teaspoon chopped fresh chilli
2 teaspoons tomato paste (purée)
2 chicken stock cubes, crumbled
1 teaspoon sesame oil
450 g (1 lb) can chopped
   pineapple pieces
3 spring onions (scallions),
   sliced diagonally

➤ COOK NOODLES in large pan boiling water for 3 minutes; drain well.

1 Slice corn diagonally. Heat oil in wok; add capsicum, celery, carrot and mushrooms. Stir over high heat for 5 minutes.

2 Add corn and noodles. Reduce heat to low; cook 2 minutes. Blend cornflour with vinegar in small mixing bowl until smooth. Add chilli, tomato paste, stock cubes, oil and undrained pineapple, stir to combine.

3 Pour pineapple mixture over ingredients in wok. Stir over medium heat 5 minutes, or until mixture boils and sauce thickens. Add spring onions; serve immediately.

COOK'S FILE

Variation: Thinly sliced Chinese barbecued pork (char siu) can be added to this dish if desired.

1

2

3

STUFFED BABY PUMPKINS

Preparation time: 15 minutes
Total cooking time: 50 minutes
Makes 4

4 medium golden nugget
  pumpkins
95 g (1/2 cup) boiled rice
2 teaspoons curry paste
1 tablespoon finely chopped
  coriander (cilantro)
1 green apple, finely chopped
1 small zucchini (courgette),
  finely chopped
1 small carrot, finely chopped

60 g (2¼ oz) button mushrooms,
  thinly sliced
100 g (3½ oz) bunch asparagus,
  chopped
2 teaspoons currants
¼ teaspoon garam masala
60 g (2¼ oz) butter, melted

➤ PREHEAT OVEN to 210°C (415°F/Gas 6–7). Cut top off each pumpkin; set aside. Scoop out seeds and discard.

1 Arrange the pumpkins in a medium ovenproof dish, then replace tops. Add 60 ml (¼ cup) water to the dish and cover firmly with foil. Bake 30 minutes. Remove the pumpkins from the dish. Drain the water and grease the dish with melted butter or oil.

2 Combine the rice, curry paste, coriander, apple, zucchini, carrot, mushrooms, asparagus, currants, garam masala and butter in a bowl, then mix well.

3 Spoon the rice and vegetable mixture into the cavity of each pumpkin. Top each pumpkin with lid. Cover the dish with foil and bake, covered, for 20 minutes, or until the vegetables are just cooked.

COOK'S FILE

Storage time: Cook this dish just before serving.

Hint: To give the filled pumpkins a shiny appearance, brush the shell and lid with oil before serving.

1

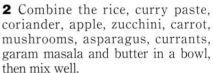
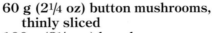

2

3

CARAMELIZED ONION AND SPINACH TART

Preparation time: 25 minutes +
  10 minutes standing
Total cooking time: 2 hours
Serves 6

125 g (1 cup) plain (all-purpose)
  flour
90 g (3¼ oz) butter, chopped
1 egg yolk

Filling
5 medium (1 kg/2 lb 4 oz)
  onions, thinly sliced
45 g (¼ cup) soft brown sugar
2 tablespoons brown vinegar
1 bay leaf
3 dried chillies

Topping
130 g (2 cups) finely chopped
  English spinach
125 g (1 cup) grated Cheddar
  cheese
30 g (¼ cup) self-raising flour
1 teaspoon mustard powder
310 ml (1¼ cups) cream
2 eggs, lightly beaten

➤ PREHEAT OVEN to 210°C (415°F/ Gas 6–7). Grease a 20 cm (8 inch) round springform tin.
1 Place flour and butter in food processor bowl. Using the pulse action, process for 15 seconds, or until the mixture is fine and crumbly. Add egg yolk and 1 tablespoon water and process for 20 seconds until smooth.
2 Press pastry evenly over base of prepared tin; refrigerate 10 minutes. Cut a sheet of greaseproof paper large enough to cover the pastry-lined tin. Spread a layer of rice evenly over paper. Bake 20 minutes; remove from oven and discard rice. Allow to cool.

Combine onions, sugar, vinegar, bay leaf, chillies and 185 ml (¾ cup) water in medium heavy-based pan. Stir over medium heat until sugar has dissolved and mixture is boiling. Reduce the heat and simmer, covered, 1 hour, stirring occasionally. Drain any excess liquid; discard bay leaf and chillies.
3 Spread the cooled onion mixture evenly over the pastry base. Spoon the topping over the onion. Bake for

45 minutes, or until topping is golden. Serve warm or cool.
To make Spinach Topping: Place all ingredients in large mixing bowl and mix well.

COOK'S FILE

Storage time: This dish can be made up to one day ahead. Store in refrigerator. Reheat gently before serving.

1

3

SPICY WINTER CASEROLE

Preparation time: 35 minutes +
  overnight soaking
Total cooking time: 1 hour 15 minutes
Serves 6

1 small eggplant (aubergine),
  cut into 2 cm (³/4 inch) cubes
1 tablespoon salt
220 g (1 cup) dried chickpeas
2 tablespoons olive oil
2 onions, sliced
2 garlic cloves, crushed
2 tablespoons grated fresh ginger
1 tablespoon ground cumin
2 teaspoons paprika
¹/4 teaspoon saffron threads

1¹/2 teaspoons chilli powder
1.5 litres (6 cups) vegetable
  stock
2 carrots, thinly sliced
2 turnips, cut into 2 cm (³/4 inch)
  cubes
3 zucchini (courgettes), cut into
  2 cm (³/4 inch) slices
300 g (10¹/2 oz) pumpkin, cut
  into 3 cm (1¹/4 inch) cubes
2 tomatoes, chopped
10 g (¹/3 cup) chopped flat-leaf
  (Italian) parsley

➤ SPREAD EGGPLANT out in a single layer and sprinkle generously with salt. Stand for 20 minutes, rinse well and pat dry with paper towels.
1 Place chickpeas in a medium bowl, cover with water and soak overnight.

2 Heat oil in a large pan. Cook onion over medium heat for 5 minutes until golden, stirring occasionally. Add garlic, ginger and spices, stir-fry for 1 minute. Drain the chickpeas and add to the pan with the stock. Bring to the boil, reduce the heat, cover and simmer for 40 minutes, until chickpeas are just tender. Stir occasionally.
3 Add the carrots and turnips to pan, simmer 15 minutes. Add remaining vegetables. Simmer, covered, for 15 minutes, or until vegetables are tender, stirring occasionally. Stir in parsley and serve with rice.

COOK'S FILE

Storage time: Casserole may be made up to two days in advance. Store in refrigerator.

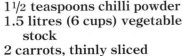

HERBED POTATO BAKE

Preparation time: 12 minutes
Total cooking time: 1 hour
Serves 4–6

6 medium potatoes
2 red onions
2 large zucchini (courgettes)
2 spring onions (scallions)
100 g (3¹/2 oz) salami (optional)
310 g (1¹/4 cups) sour cream
2 garlic cloves, crushed

2 tablespoons chopped fresh
  parsley
2 tablespoons chopped fresh
  chives
25 g (¹/4 cup) dry breadcrumbs

➤ PREHEAT OVEN to 180°C (350°F/ Gas 4). Brush a large, shallow ovenproof dish with melted butter.
1 Using a sharp knife, thinly slice the potatoes, onions, zucchini, spring onions and salami.
2 Combine the sliced vegetables and salami with the cream, garlic and

herbs in a large bowl. Season with salt and freshly ground black pepper, then mix well. Transfer the mixture to the prepared dish and smooth the surface. Cover with foil. Bake for 20 minutes, then remove the foil.
3 Sprinkle breadcrumbs over the vegetable mixture; bake on top shelf of oven for 40 minutes, or until golden and potatoes are tender.

COOK'S FILE

Storage time: Cook this dish just before serving.

*Spicy Winter Casserole (top)
and Herbed Potato Bake.*

LEEK AND TURNIP PIE

Preparation time: 45 minutes
Total cooking time: 1 hour +
    12 minutes
Makes one 25 cm (10 inch) pie

250 g (2 cups) plain (all-purpose)
    flour
125 g (4¹/2 oz) butter, chopped
50 g (¹/2 cup) grated Parmesan
    cheese
1–2 tablespoons iced water

Filling
2 medium leeks
100 g (3¹/2 oz) butter
750 g (1 lb 10 oz) white turnips,
    washed, peeled, thinly sliced
2 tablespoons caraway seeds
2 tablespoons soft brown sugar
2 tablespoons red wine vinegar
15 g (¹/4 cup) chopped fresh
    basil
2 tablespoons plain (all-purpose)
    flour
125 g (4¹/2 oz) Cheddar cheese,
    grated
25 g (¹/4 cup) grated Parmesan
    cheese
1 egg, lightly beaten

➤ PREHEAT OVEN to 180°C (350°F/
Gas 4). Brush a 25 cm (10 inch) pie
plate with melted butter or oil.
1 Sift flour into a large bowl and add
butter. Using fingertips, rub butter
into flour until mixture is fine and
crumbly. Add cheese and almost all
the water; mix to a firm dough,
adding a little more water if necessary. Turn onto a lightly floured surface, knead 2 minutes, or until smooth.
2 Divide dough in two. Roll out one
portion between two sheets of baking
paper, large enough to cover the base
and sides of pie plate. Trim edges. Cut
a sheet of greaseproof paper large
enough to cover pastry-lined tin.
Spread a layer of dried beans evenly
over paper. Bake 8 minutes. Remove
from oven; discard paper and beans.
Return pastry to oven for 5 minutes,
or until lightly golden. Cool.
3 Slice leeks finely. Heat butter in
large pan, add turnips and leeks. Cook
over medium heat for 4 minutes, or
until coated with butter. Cover, cook
for 10 minutes, shaking pan occasionally to prevent sticking. Add caraway
seeds and brown sugar. Stir until
sugar melts. Add vinegar, cook for
1 minute. Remove from heat, cool
slightly. Stir through basil and

1 tablespoon of the flour. Season.
4 Spoon one-third of turnip mixture
over pastry. Combine remaining flour
and cheeses. Sprinkle one-third over
turnip mixture. Continue layering, finishing with cheese. Roll remaining
pastry into 4 cm (1¹/2 inch) diameter
log, cut into 5 mm (¹/4 inch) slices.
Place overlapping pastry circles
around edge of pie. Brush between
each round with egg. Bake 30–40 minutes, or until golden.

VEGETABLE PILAF

Preparation time: 20 minutes
Total cooking time: 35 minutes
Serves 4

60 ml (1/4 cup) olive oil
1 medium onion, sliced
2 garlic cloves, crushed
2 teaspoons ground cumin
2 teaspoons paprika
1/2 teaspoon allspice
300 g (1 1/2 cups) long-grain
  rice

375 ml (1 1/2 cups) vegetable
  stock
185 ml (3/4 cup) white wine
3 medium tomatoes, chopped
150 g (5 1/2 oz) button
  mushrooms, sliced
2 medium zucchini (courgettes),
  sliced
150 g (5 1/2 oz) broccoli, cut into
  florets

➤ HEAT OIL in a heavy-based pan.
1 Add the onion and cook for 10 minutes over medium heat until golden brown. Add the garlic and spices,

cook 1 minute until aromatic.
2 Add rice to the pan, stir until well combined. Add vegetable stock, wine, tomatoes and mushrooms, bring to the boil. Reduce heat to low, cover pan with tight-fitting lid. Simmer for 15 minutes.
3 Add zucchini and broccoli to the pan, replace the lid and cook further 5–7 minutes, until vegetables are just tender. Serve immediately.

COOK'S FILE

Storage time: Cook this dish just before serving.

SIDE DISHES

SUGAR PEAS AND CARROTS IN LIME BUTTER

Preparation time: 10 minutes
Total cooking time: 10 minutes
Serves 4

125 g (4½ oz) carrots
125 g (4½ oz) sugar snap peas
50 g (1¾ oz) butter
2 garlic cloves, crushed
1 tablespoon lime juice
½ teaspoon soft brown sugar
1 lime

➤ PEEL CARROTS and cut into thin diagonal slices.

1 Wash and string sugar snap peas. Heat butter in a large heavy-based frying pan. Add garlic, cook over low heat for 1 minute. Add juice and sugar. Cook, stirring over heat until sugar has completely dissolved.

2 Add peas and carrots, cook over medium heat 2–3 minutes, or until tender but still crisp. Serve hot. Garnish with lime zest.

3 To make lime zest: Peel lime rind into long strips using a vegetable peeler. Remove all white pith. Cut into long thin strips with a sharp knife.

PUMPKIN WITH CHILLI AND AVOCADO

Preparation time: 20 minutes
Total cooking time: 10 minutes
Serves 6

700 g (1 lb 9 oz) pumpkin
2 tablespoons olive oil
1 tablespoon chopped fresh
   coriander (cilantro) leaves
1 tablespoon chopped fresh mint
2 teaspoons sweet chilli sauce
1 small red onion, finely
   chopped
2 teaspoons balsamic vinegar
1 teaspoon soft brown sugar
1 large avocado

➤ SCRAPE SEEDS from inside of pumpkin.

1 Cut pumpkin into thin slices. Remove skin. Cook in a large pan of simmering water until tender but still firm. Remove from heat; drain well.

2 Combine oil, coriander, mint, chilli, onion, vinegar and sugar in a small bowl. Mix well. Cut avocado in half. Remove stone using a sharp-bladed knife. Peel skin from avocado. Discard. Cut avocado in thin slices.

3 Combine warm pumpkin and avocado in a serving bowl. Gently toss coriander dressing through. Serve immediately.

COOK'S FILE

Storage time: Assemble this dish just before serving. The coriander dressing can be made up several hours in advance. Store, covered, in the refrigerator.

Variation: Add one small red chilli, finely chopped, to dressing if liked. For a milder flavour, remove the seeds and membranes of the chilli.

1

2

3

EGGPLANT AND TOMATO CAPONATA

Preparation time: 25 minutes
Total cooking time: 15 minutes
Serves 4

500 g (1 lb 2 oz) small eggplant
  (aubergines)
2 tablespoons salt
1 onion
3 stalks celery
2 tomatoes
125 ml (1/2 cup) olive oil
1 tablespoon capers, lightly
  crushed
1 tablespoon caster (superfine)
  sugar
2 tablespoons white wine vinegar

➤ SLICE EGGPLANT in half length-ways. Cut in thin slices. Place on a flat tray. Sprinkle with the salt and leave for 20 minutes.

1 Cut onion in thin slices. Cut celery into thin slices diagonally. Mark a small cross in the base of each tomato. Place in boiling water for 1–2 minutes, then plunge into cold water. Peel the skin down from the cross. Cut the tomato into wedges and squeeze gently to remove seeds.

2 Heat 1 tablespoon oil in a large frying pan. Add the onion and cook over medium heat 2 minutes, or until soft and slightly golden. Add celery, cook 2 minutes. Add the tomatoes, capers, sugar and vinegar. Season to taste. Cook for 5 minutes. Remove from heat and set aside.

3 Rinse the eggplant, drain. Pat dry with paper towels. Heat remaining oil in a medium frying pan, add eggplant. Cook over medium heat for 4–5 minutes, or until soft and golden brown. Drain eggplant on paper towels. Add eggplant to onion mixture. Stir until well combined. Leave to cool. Serve at room temperature.

COOK'S FILE

Storage time: This dish can be made two days ahead. Store in an air-tight container in refrigerator. Bring to room temperature to serve.

Hints: Sliced green olives can be added to this dish if desired. Add them with the capers.

Caponata goes well with grilled fish such as tuna, or grilled steak.

TWO-POTATO HASH BROWNS

Preparation time: 20 minutes
Total cooking time: 25 minutes
Serves 4–6

3 rashers bacon, finely chopped
    (optional)
500 g (1 lb 2 oz) potatoes
250 g (9 oz) orange sweet potato
1 large onion, finely chopped
2 tablespoons olive oil
sour cream, for serving
2 tablespoons chopped fresh
    chives

➤ PREHEAT OVEN to 180°C (350°F/ Gas 4). Brush a baking tray with melted butter or oil. Grease an egg ring to use as a mould for hash browns.
1 Place bacon in small pan. Cook over medium-high heat 3 minutes. Drain on paper towels. Chop finely. Peel and grate the potato and sweet potato. Place the combined potatoes in muslin and squeeze out any excess moisture.
2 Place the potatoes, onion, bacon and oil in a bowl. Season to taste with salt and freshly ground black pepper. Toss well to ensure even mixing. Press spoonfuls of mixture into the egg ring on a tray. Level surface.

Remove the ring and repeat with remaining mixture until all the mixture is shaped.
3 Bake for 20 minutes, or until crisp and golden. Serve immediately, topped with sour cream and chives.

COOK'S FILE

Storage time: Combined raw mixture can be prepared 12 hours ahead. Store, covered, in the refrigerator.
Note: Hash browns can be pan-fried in a small amount of oil. To obtain a crisp result, it is essential that the moisture is squeezed out of the potato before mixing with other ingredients.

1

2

3

SPRING ONION AND CELERY BUNDLES

Preparation time: 20 minutes
Total cooking time: 10 minutes
Serves 6

4 stalks celery
1 bunch spring onions (scallions)
30 g (1 oz) butter
1 teaspoon celery seeds
1 tablespoon honey
125 ml (1/2 cup) chicken stock
1 teaspoon soy sauce
1 teaspoon cornflour
   (cornstarch)

➤ CUT CELERY into 10 cm (4 inch) lengths, then into strips the same thickness as spring onions.

1 Cut root from spring onions. Cut into 10 cm (4 inch) lengths. Reserve spring onion tops for ties. Plunge spring onion tops into boiling water 30 seconds, or until bright green, then plunge immediately into iced water. Drain and pat dry with paper towels.

2 Combine spring onion and celery stalks. Divide into six bundles. Tie each bundle firmly with a spring onion top.

3 Heat butter in frying pan. Fry bundles quickly over medium-high heat 1 minute each side. Remove from pan.

Add celery seeds, cook 30 seconds. Add honey, stock, soy sauce and blended cornflour and 1 teaspoon water. Bring to boil, reduce heat, stirring continuously. Add spring onion and celery bundles. Simmer gently 7 minutes, or until bundles are just tender. Serve with a little of the cooking liquid.

COOK'S FILE

Storage time: Bundles can be assembled up to 12 hours ahead. Cover with a damp tea towel and store in the refrigerator. Cook just before serving.
Hint: This side dish is suitable to accompany meat, fish or poultry dishes. Use beef stock in place of chicken.

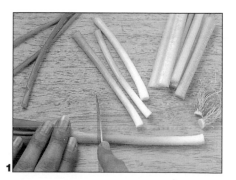

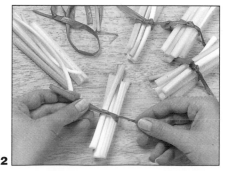

SPROUT AND PEAR SALAD WITH SESAME DRESSING

Preparation time: 30 minutes +
  50 minutes refrigeration
Total cooking time: Nil
Serves 6

1 punnet snow pea sprouts
250 g (9 oz) fresh bean sprouts
1 bunch chives
100 g (3¹/2 oz) snow peas
  (mangetout)
1 stalk celery
2 firm pears, not green

15 g (¹/2 cup) coriander (cilantro)
  sprigs
black or white sesame seeds,
  for garnish

Sesame Dressing
2 tablespoons soy sauce
1 teaspoon sesame oil
1 tablespoon soft brown sugar
2 tablespoons peanut oil
1 teaspoon rice vinegar

➤ WASH snow pea sprouts; drain.
1 Remove brown tip from bean sprouts. Cut chives into 4 cm (1¹/2 inch) lengths. Cut snow peas and celery into thin matchstick strips.

2 Peel and core pears. Cut into thin strips slightly wider than celery and snow peas. Cover with water to prevent discolouring.
To make Sesame Dressing: Combine all ingredients in small screwtop jar and shake well.
3 Drain pears. Combine the pears, snow pea and bean sprouts, snow peas, celery and coriander in large serving bowl. Pour over dressing, toss lightly to combine. Sprinkle with sesame seeds. Serve immediately.

COOK'S FILE

Storage time: Dressing can be prepared two days ahead. Refrigerate.

STIR-FRIED CHINESE VEGETABLES

Preparation time: 15 minutes
Total cooking time: 7 minutes
Serves 4

300 g (10¹/₂ oz) baby bok choy
    (pak choi)
100 g (3¹/₂ oz) snake beans
2 spring onions (scallions)
150 g (5¹/₂ oz) broccoli
1 red capsicum (pepper)
2 tablespoons oil
2 garlic cloves, crushed
2 teaspoons grated fresh ginger
1 tablespoon sesame oil
2 teaspoons soy sauce

➤ WASH AND TRIM thick stalks
from bok choy.
1 Cut leaves into wide strips. Cut
snake beans into 5 cm (2 inch) lengths.
Slice spring onions diagonally. Cut
broccoli into small florets. Cut the
capsicum into diamonds about 2 cm
(³/₄ inch) wide.
2 Heat oil in large heavy-based frying pan or wok. Add garlic and ginger
and cook over medium heat for 30 seconds, stirring constantly. Add beans,
spring onions and broccoli, stir-fry for
3 minutes.
3 Add capsicum, stir-fry a further
2 minutes; add bok choy and stir-fry
for 1 minute. Toss through sesame oil
and soy sauce. Transfer vegetables to
a serving dish and serve immediately.

COOK'S FILE

Storage time: Cook this dish just
before serving.
Hints: It is important not to overcook
vegetables when stir-frying. Use the
minimum amount of oil and cook over
medium-high heat, stirring and toss-
ing constantly. They will soften
slightly but should never be cooked to
a limp and greasy state. Add leafy
green vegetables last, and cook only
until the leaves have just softened.
Cutting vegetables into thin, even-
sized pieces and on the diagonal helps
them to cook quickly.

Variations: Use any Chinese vegetable in this dish. Add a little hoisin
sauce at the end of cooking if desired.
Note: Sesame oil is available at
supermarkets and Chinese food stores.
It is usually added at the end of cooking and used sparingly.

1

2

3

SAVOURY FENNEL CRUMBLE

Preparation time: 20 minutes
Total cooking time: 40 minutes
Serves 6

2 fennel bulbs
60 ml (¼ cup) lemon juice
2 tablespoons lemon juice, extra
1 tablespoon honey
1 tablespoon plain (all-purpose) flour
310 ml (1¼ cups) cream

Crumble Topping
75 g (¾ cup) rolled oats
60 g (½ cup) plain (all-purpose) flour

110 g (1 cup) black rye breadcrumbs, made from 3 slices bread
60 g (2¼ oz) butter
1 garlic clove, crushed

➤ PREHEAT OVEN to 180°C (350°F/ Gas 4). Grease a 2 litre (8-cup) oven-proof serving dish.

1 Trim the fennel and cut into thin slices. Wash and drain well. Bring a large pan of water to the boil. Add lemon juice and fennel slices. Cook over medium heat for 3 minutes. Drain and rinse under cold water.

2 Place fennel in a bowl. Add extra juice and honey. Season with pepper; toss to combine. Sprinkle with flour. Spoon into dish; pour cream over.

3 To make Crumble Topping:

Combine oats, flour and breadcrumbs. Heat butter in small pan. Add garlic, cook 30 seconds. Pour over dry ingredients and mix well. Sprinkle mixture over fennel. Bake 20–30 minutes, or until fennel is tender and crumble is browned.

COOK'S FILE

Storage time: This dish can be assembled up to eight hours prior to cooking. Store, covered, in the refrigerator. Bring to room temperature before baking.
Hints: White or wholemeal bread-crumbs can be used in place of rye bread. Fennel has an aniseed flavour. Blanching it before use softens the texture slightly and reduces the strong flavour.

ALMOND AND BROCCOLI STIR-FRY

Preparation time: 5 minutes
Total cooking time: 5 minutes
Serves 4

1 teaspoon coriander seeds
500 g (1 lb 2 oz) broccoli
60 ml (¼ cup) olive oil
2 tablespoons slivered almonds
1 garlic clove, crushed

1 teaspoon finely shredded fresh ginger
2 tablespoons red wine vinegar
1 tablespoon soy sauce
2 teaspoons sesame oil
1 teaspoon toasted sesame seeds

➤ PUT CORIANDER SEEDS in a plastic bag, hit with a rolling pin to crack.

1 Cut the broccoli into small florets.
2 Heat oil in wok or large heavy-based frying pan. Add coriander seeds

and almonds. Stir quickly over medium heat for 1 minute, or until the almonds are golden.
3 Add garlic, ginger and broccoli to pan. Stir-fry over high heat 2 minutes. Remove pan from heat. Pour combined vinegar, soy sauce and oil into wok. Toss until broccoli is well coated. Serve warm or cold, sprinkled with sesame seeds.

COOK'S FILE

Storage time: This dish may be prepared two hours ahead of serving.

Savoury Fennel Crumble (top) and Almond and Broccoli Stir-fry.

MEDITERRANEAN-STYLE BRAISED LETTUCE

Preparation time: 5 minutes
Total cooking time: 3 minutes
Serves 4

12 large cos (romaine) lettuce
   leaves
80 ml (1/3 cup) olive oil
1/2 red capsicum (pepper), cut
   into fine matchstick strips
2 spring onions (scallions), cut
   into 1 cm (1/2 inch) pieces
1 tablespoon chopped fresh
   chives
1 tablespoon lemon juice
2 teaspoons crumbled feta
   cheese
1/4 teaspoon cracked black
   pepper

➤ TEAR EACH LETTUCE leaf into
four pieces. Heat oil in medium pan;
add capsicum. Stir over low heat for
1 minute.

1 Add lettuce to pan, toss over high
heat 1 minute, or until leaves are well
coated with oil. Remove lettuce from
pan. Reduce heat to low.

2 Add spring onions, chives and juice
to pan; cook, covered, for 30 seconds.
Remove pan from heat.

3 Combine lettuce and spring onion
mixture. Place on serving plate.
Sprinkle with feta and black pepper.
Serve warm or cold.

COOK'S FILE

Notes: The beauty of this dish is in the
quick cooking. It is important to have
all the ingredients on hand before
beginning to cook, as prolonged cooking causes the lettuce to develop a bitter flavour and lose colour.
Serve with barbecued fish or chicken.

1

2

3

BARBECUED MARINATED VEGETABLES

Preparation time: 20 minutes +
  1 hour marinating
Total cooking time: 10 minutes
Serves 4–6

3 slender eggplant (aubergines)
3 zucchini (courgettes)
1 red capsicum (pepper)
1 green capsicum (pepper)
1 red onion

2 garlic cloves, crushed
2 teaspoons finely chopped
  fresh basil leaves
2 teaspoons chopped fresh
  thyme leaves
60 ml (¼ cup) olive oil
2 tablespoons balsamic vinegar

➤ CUT EGGPLANT and zucchini
into 12 cm (5 inch) diagonal slices.
1 Cut the capsicum into strips 2 cm
(¾ inch) wide, and cut the onion into
8 wedges.
2 Place all vegetables into a large

bowl. Add the garlic, herbs, oil and
vinegar, toss lightly to combine. Cover
with plastic wrap; let stand for 1 hour.
3 Preheat barbecue (grill). Drain vegetables from marinade. Place on a
lightly greased grill or flat plate. Cook
over medium-high heat 8–10 minutes,
or until tender and slightly charred.
Serve with barbecued meat.

C O O K ' S   F I L E

Storage time: Vegetables may be
prepared and cooked up to one hour
ahead. Serve at room temperature.

1

2

3

CHILLI SWEET POTATO AND EGGPLANT CRISPS

Preparation time: 5 minutes
Total cooking time: 20 minutes
Serves 4–6

1 orange sweet potato
   (300 g/10½ oz)
1 slender eggplant (aubergine)
   (350 g/12 oz)
oil, for deep-frying

¼ teaspoon ground chilli powder
¼ teaspoon ground coriander
1 teaspoon chicken salt

➤ PEEL sweet potato.
1 Cut sweet potato and eggplant into long thin strips, similar in size. Place in a large bowl, mix.
2 Heat oil in a deep heavy-based pan. Gently lower half the combined sweet potato and eggplant into the moderately hot oil. Cook over medium-high heat for 10 minutes, or until golden and crisp. Carefully remove the crisps

from the oil with tongs or slotted spoon. Drain on paper towels. Repeat cooking process with remaining sweet potato and eggplant.
3 Combine chilli, coriander and salt. Sprinkle over the hot crisps and toss until well coated. Serve immediately.

COOK'S FILE

Storage time: Cook this dish just before serving.
Hint: These tasty crisps go well with casual dishes, such as grilled chicken breasts on rolls with salad.

ZUCCHINI WITH CUMIN CREAM

Preparation time: 5 minutes
Total cooking time: 12 minutes
Serves 4–6

1 lemon
4 large zucchini (courgettes)
30 g (1 oz) butter
1 tablespoon oil
1/2 teaspoon cumin seeds
125 ml (1/2 cup) cream

► GRATE ZEST from lemon to make 1/2 teaspoon and squeeze 2 teaspoons lemon juice. Set aside.

1 Cut zucchini into diagonal slices 5 mm (1/4 inch) thick. Heat the butter and oil in a large frying pan; add half the zucchini. Cook over medium-high heat for 2 minutes each side, or until golden. Remove from pan; drain on paper towels and keep warm. Repeat with remaining zucchini.

2 Add cumin seeds to pan, stir over low heat for 1 minute. Add lemon zest and juice; bring to the boil. Add cream to pan, boil for 2 minutes, or until sauce thickens slightly (see Note); season to taste with salt and pepper.

3 Return zucchini to pan. Stir over low heat 1 minute, or until just heated through. Serve warm or cold.

COOK'S FILE

Storage time: Cook this dish up to one hour before serving. Refrigerate until ready to serve.

Note: Do not boil sauce for too long as it may separate.

1

2

3

POTATO AND POPPYSEED FOCACCIA

Preparation time: 30 minutes +
  1 hour rising
Total cooking time: 40 minutes
Serves 4

250 g (2 cups) plain (all-purpose)
  flour
1 tablespoon dried yeast
1 teaspoon sugar
1 teaspoon salt
310 g (1¹/₃ cups) mashed potato
60 ml (¹/₄ cup) warm water
80 ml (¹/₃ cup) olive oil
2 tablespoons semolina
1 egg, lightly beaten
2 teaspoons poppyseeds

➤ SIFT FLOUR into mixing bowl.
1 Add yeast, sugar, salt and mashed potato. Pour in combined water and oil. Using a knife or spatula, mix to a soft dough.
2 Turn dough onto a well-floured surface, knead for 10 minutes, until smooth. Sprinkle surface with more flour as necessary. Shape into a ball, place into a large, lightly oiled mixing bowl. Leave, covered with plastic wrap, in a warm place for 30 minutes, until well risen.
3 Preheat oven to 180°C (350°F/Gas 4). Knead dough again for 5 minutes. Sprinkle the base of a greased 23 cm (9 inch) square cake tin with semolina. Press dough into tin with floured hands. With a skewer, prick four rows of four holes into dough. Cover with plastic wrap and leave in a warm place for 20 minutes, until well risen. Brush lightly with the beaten egg; sprinkle with poppyseeds. Bake 40 minutes, until golden brown. Turn focaccia onto a wire rack to cool.

COOK'S FILE

Storage time: Focaccia is best served on the day it is made, however, it is delicious split and toasted when one or two days old.
Hint: The potato adds moisture to the dough.
Variations: This recipe is open to many variations in flavour. Fresh or dried herbs may be kneaded into the dough. Try sprinkling with sesame seeds or sliced olives instead of poppyseeds.

ASPARAGUS AND SNOW PEA SALAD

Preparation time: 15 minutes
Total cooking time: 5 minutes
Serves 4–6

200 g (7 oz) snow peas
  (mangetout)
1 bunch asparagus

Dressing
2 tablespoons peanut oil
3 teaspoons sesame oil
3 teaspoons rice vinegar or red
  wine vinegar
¹/₂ teaspoon sugar
1 tablespoon sesame seeds

➤ TOP AND TAIL snow peas.
1 Trim any woody ends from asparagus. Cut spears diagonally in half. Place in pan of boiling water. Stand for 1 minute, then drain and plunge into iced water. Drain well.
2 **To make Dressing:** Place oils, vinegar and sugar in a small screwtop jar; shake well. Place asparagus and snow peas in a serving bowl. Pour dressing over; toss to combine.
3 Cook sesame seeds in a dry frying pan over medium heat for 1–2 minutes, until lightly golden, sprinkle over salad. Serve immediately.

COOK'S FILE

Storage time: Prepare vegetables up to four hours in advance; add dressing up to one hour in advance.
Hint: Rice vinegar is available at Chinese food stores.

*Potato and Poppyseed Focaccia (top)
and Asparagus and Snow Pea Salad.*

WARM TOMATO AND HERB SALAD

Preparation time: 20 minutes
Total cooking time: 15 minutes
Serves 6

1 garlic clove, crushed
1 tablespoon olive oil
French bread, cut into 12 slices, 2 cm (3/4 inch) thick
1 tablespoon olive oil, extra
250 g (9 oz) punnet cherry tomatoes
250 g (9 oz) punnet yellow pear tomatoes

15 g (1/4 cup) shredded basil leaves
1 tablespoon chopped fresh tarragon
15 g (1/4 cup) chopped fresh parsley

➤ PREHEAT OVEN to 180°C (350°F/ Gas 4).

1 Combine garlic and oil in a small bowl. Brush one side of bread lightly with the oil, place on a baking tray. Bake for 7 minutes; turn over, brush other side and bake 5 minutes. Cool.

2 Heat oil in a frying pan. Add whole tomatoes and stir-fry over medium heat for 2 minutes, until just soft.

3 Add herbs to tomatoes, stir-fry for 1 minute, or until combined. Serve warm with the croûtons.

COOK'S FILE

Storage time: Croûtons can be made up to eight hours in advance, cooled completely and stored in an airtight container. Tomatoes should be cooked just before serving.

Hints: Yellow pear tomatoes are similar in size to cherry tomatoes and sold in punnets. If unavailable, use all cherry tomatoes. Miniature tomatoes are easy to grow. They are more pest-resistant than large tomatoes and can even be grown in a pot on a sunny balcony.

1

2

3

HONEYED BABY TURNIPS WITH LEMON THYME

Preparation time: 10 minutes
Total cooking time: 7 minutes
Serves 4

500 g (1 lb 2 oz) baby turnips
40 g (1¹/2 oz) butter
60 g (¹/4 cup) honey
3 teaspoons lemon juice
¹/2 teaspoon grated lemon zest
3 teaspoons chopped fresh
    lemon thyme leaves

➤ RINSE AND lightly scrub turnips under water.
1 Trim tips and stalks. Cook in pan of boiling water 1 minute. Drain, refresh under cold water, drain well.
2 Heat butter in a medium pan; add honey. Bring mixture to the boil, add lemon juice and zest. Boil over high heat for 3 minutes. Add turnips to honey and lemon mixture in pan. Cook over high heat for 3 minutes, or until the turnips are almost tender and well glazed. (Test turnips with a skewer.)
3 Add lemon thyme. Remove pan from heat. Toss until turnips are well coated. Serve warm.

COOK'S FILE

Storage time: Cook this dish up to one hour before serving.
Variation: Baby carrots or baby beetroot can be used in this recipe.

1

2

3

SPICED BAKED BEETROOT

Preparation time: 15 minutes
Total cooking time: 1 hour 25 minutes
Serves 6

12 small beetroot
    (1.2 kg/2 lb 11 oz)
2 tablespoons olive oil
1 teaspoon ground cumin
1 teaspoon ground coriander
1/2 teaspoon ground cardamom
1/2 teaspoon nutmeg
3 teaspoons sugar

1 tablespoon red wine vinegar

➤ PREHEAT OVEN to 180°C (350°F/ Gas 4). Grease a baking tray.
1 Trim leafy tops from beetroot and wash thoroughly. Place on prepared tray and bake for 1 hour 15 minutes, until very tender. Cool slightly. Peel beetroot; trim tops and tails to neaten.
2 Heat oil in a large pan. Add spices and cook for 1 minute, stirring constantly, over medium heat. Add sugar and vinegar, stir for 2–3 minutes, or until sugar dissolves.
3 Add the beetroot to the pan, reduce the heat to low and stir gently for

5 minutes, until the beetroot is well glazed. Serve warm or cold.

COOK'S FILE

Storage time: This dish can be cooked up to two days ahead. Store in a covered container in the refrigerator.
Hint: This dish is ideal for picnic fare. Serve Spiced Baked Beetroot with cold or hot roast meats or poultry.
Variation: Small potatoes, sliced sweet potatoes, peeled small onions or even Brussels sprouts can be cooked in this way. Bake until soft and glaze as in Step 3. Brussels sprouts should be lightly steamed rather than baked.

INDEX

Almond and Broccoli Stir-fry, 100
Asparagus
    and Grilled Eggplant Sandwich, 65
    and Snow Pea Salad, 106
Artichokes with Tarragon
    Mayonnaise, 67
Avocado
    and Pumpkin with Chilli, 94
    Salad, Hot, 51
    Salsa, 27

Basil
    Spinach Salad, 39
    and Tomato, with Peas, 47
Bean
    Bundles, 41
    and Walnut Salad, 41
Beans
    and Cabbage, 35
    and Cashews, 40
    Garlic and Basil, 40
    in Herb Cream Sauce, 41
    Hollandaise, 40
    and Minted Tomato, 41
    Pepper, and Ham, 40
Beetroot, Spiced Baked, 110
Broccoli
    and Almond Stir-fry, 100
    with Bacon and Pine Nuts, 32
    Buttered, and Herbs 33
    with Cashews, 32
    with Cheese Sauce, 32
    Lemon, 33
    and Mushrooms, 33
    with Mustard Butter, 33
    and Onion Stir-fry, 32

Cabbage
    and Beans, 35
    Garlic Pepper, 35
    and Potato Cakes, 35
    Sautéed, 34
    and Spring Onion Stir-fry, 34
    Sweet Chilli, 35
    Sweet Red, with Caraway Seeds, 34
    Turnovers, Braised, 82
Capsicum
    Frittata, 76
    Relish, with Eggplant and
        Zucchini Pots, 56
    Soup, Red, 66
Carrot
    Pesto Slice, 75
    Ribbons, 29
    Sticks, Herbed, 29
Carrots
    Baby, Garlic Buttered, 29
    Honey-glazed, 29
    and Sugar Peas in Lime Butter, 93
Casserole, Spicy Winter, 86
Cauliflower
    with Bacon, 44
    Cheese, 44
    Fritters with Tomato Relish, 16
    au Gratin, 44
    Hot chilli, 45
    with Lime Butter, 45
    Parmesan, 45
    Spiced, 44
    with Tomato Sauce, 45
Celery and Spring Onion Bundles, 97
Cheese and Olive Slice, 13
Cheese, Two-, Risotto Cakes, 57
Cheese and Corn Chowder, 54

Chilli
    Puffs with Curried Vegetables,
        58
    Hot, Cauliflower, 45
    Cheese Dip with Crisp Potato
        Skins, 23
Coleslaw, Quick, 34
Corn
    and Cheese Chowder, 54
    Pancakes, Thai, with Coriander
        Mayonnaise, 17

Eggplant
    and Asparagus Sandwich,
        Grilled, 65
    and Sweet Potato Crisps, Chilli, 104
    and Tomato Caponata, 95
    and Zucchini Pots with
        Capsicum Relish, 56
Eggs Florentine, 39

Fennel
    Crumble, Savoury, 100
    Soup, Creamed, 59
Filo Vegetable Pouches, 14
Fritters, Vegetable with Tomato
    Sauce, 21

Garlic
    and Basil Beans, 40
    Buttered Baby Carrots, 29
    Onions, 42
    Pepper Cabbage, 35
Gnocchi, Pumpkin with Sage
    Butter, 72
Gourmet Vegetable Pizza, 78

Hash Browns, Two-Potato, 96

Leek and Turnip Pie, 90
Lemon Broccoli, 33
Lettuce, Braised Mediterranean-
    style, 102

Mexican-style Vegetables, 69
Mushroom
    Caps with Garlic and Thyme, 62
    and Bean Sauce, Creamy, with
        Fettuccine, 70
Mushrooms and Broccoli, 33

Olive and Cheese Slice, 13
Olive, Bites and Spinach, 24
Onion
    Caramelized, and Spinach
        Tart, 85
    Rings, Curried, 42
    Salsa, Quick, 43
    Soup, Rich Red, 60
Onions
    Baked, 43
    Caramelized, 43
    Garlic, 42
    Golden Baby, 42
    Golden and Peas, 47
    Spicy, and Tomatoes, 42
    and Thyme, 43

Parmesan Cauliflower, 45

Pea Soup, Green, 60
Pear and Sprout Salad with Sesame
    Dressing, 98
Peas
    and Bacon, 47
    with Basil and Tomato, 47
    Creamed, 47
    and Golden Onions, 47
    Minted, 46
    Peppered, and Garlic, 47
    Sautéed, and Spring Onions, 46
    Snow, and Asparagus Salad, 106
    Sugar, and Carrots in Lime
        Butter, 93
    Sweet Coriander, 46
Pilaf, Vegetable, 91
Pizza, Gourmet Vegetable, 78
Potato
    Bake, Herbed, 86
    Bake, Quick Cheesy, 31
    Cakes and Cabbage, 35
    Cakes with Prawns and
        Zucchini, 74
    Chips or Curls, 30
    Creamy, 30
    Hash Browns, Two-, 96
    and Poppyseed Focaccia, 106
    Salad, 31
    Skins, Crisp, with Chilli Cheese
        Dip, 23
Potatoes
    Duchess, 30
    Golden Roasted, 30
    Hasselback, 31
    Herbed New, 31
Pumpkin
    Baked, 36
    Candied, 36
    with Chilli and Avocado, 94
    with Chive Butter, 36
    with Garlic and Herb Butter, 37
    Gnocchi with Sage Butter, 72
    and Nutmeg Purée, 36
    Ribbons, Fried, 37
    Soup, 37
    Sweet Spiced, 37
    Sweet Spiced Golden Nuggets, 80
Pumpkins, Stuffed Baby, 84

Risotto Cakes, Two-Cheese, 57
Risotto, Vegetable, 88

Salad Baskets with Berry
    Dressing, 49
Salmon and Spinach Terrine, 53
Samosas, Vegetable, 26
Savoury Tarts, 18
Snow Pea and Asparagus
    Salad, 106
Spinach
    and Buttered Chives, 38
    and Caramelized Onion Tart, 85
    Creamed, 38
    Croquettes with Minted Yoghurt
        Sauce, 15
    Fettuccine with Rich Tomato
        Sauce, 89
    and Olive Bites, 24
    Salad, Basil, 39

Salad, Sweet Chilli, 39
    and Salmon Terrine, 53
    Shredded, and Bacon, 38
    Soufflés, Individual, 63
    and Spring Onion Salad, 39
    with Vinaigrette, 38
Spring Onion
    and Cabbage Stir-fry, 34
    and Celery Bundles, 97
    and Peas, Sautéed, 46
    and Spinach Salad, 39
Sprout and Pear Salad with Sesame
    Dressing, 98
Stir-fried
    Cabbage and Spring Onion, 34
    Chinese Vegetables, 99
    Vegetables, Quick, 72
    Stuffed Baby Pumpkins, 84
Sweet
    and Sour Noodles and
        Vegetables, 83
    Spiced Golden Nuggets, 80
Sweet Potato
    Chilli, and Eggplant Crisps, 104
    Muffins, 18
    Roast, with Coriander Pesto and
        Spring Salad, 50
    Soup, 52

Thai Corn Pancakes with
    Coriander Mayonnaise, 17
Tomato
    and Beans, Minted, 41
    and Eggplant Caponata, 95
    and Herb Salad, Warm, 108
    Relish, with Cauliflower Fritters, 16
    Sauce, Rich, with Spinach
        Fettuccine, 89
    Sauce, with Vegetable Fritters, 21
    Spicy Onions and, 42
Tomatoes, Grilled, with
    Bruschetta, 54
Turnip and Leek Pie, 90
Turnips, Honeyed Baby, with
    Lemon Thyme, 109

Vegetable
    Fritters with Tomato Sauce, 21
    Lasagne, 71
    Pie, 77
    Pilaf, 91
    Pizza, Gourmet, 78
    Pouches, Filo, 14
    Risotto, 88
    Samosas, 26
    Stir-fry, 72
    Tart, 81
    Tempura, 64
    Wontons with Chilli Sauce, 20
Vegetables
    Barbecued Marinated, 103
    Chinese, Stir-fried, 99
    Curried, with Chilli Puffs, 58
    Mexican-style, 69

Wontons, Vegetable with Chilli
    Sauce, 20

Zucchini
    and Eggplant Pots, with
        Capsicum Relish, 56
    Fingers, 22
    with Cumin Cream, 105
    and Prawns with Potato Cakes, 74

USEFUL INFORMATION

All our recipes are tested in a special test kitchen. Standard metric measuring cups and spoons are used in the development of our recipes. All cup and spoon measurements are level. We have used 60 g (2¼ oz/Grade 3) eggs in all recipes. Sizes of cans vary from manufacturer to manufacturer and between countries—use the can size closest to the one suggested in the recipe.

Conversion Guide

1 cup	= 250 ml (9 fl oz)
1 teaspoon	= 5 ml
1 Australian tablespoon	= 20 ml (4 teaspoons)
1 UK/US tablespoon	= 15 ml (3 teaspoons)

NOTE: We have used 20 ml tablespoon measures. If you are using a 15 ml tablespoon, for most recipes the difference will not be noticeable. However, for recipes using baking powder, gelatine, bicarbonate of soda, small amounts of flour and cornflour, add an extra teaspoon for each tablespoon specified.

Dry Measures

30 g	= 1 oz
250 g	= 9 oz
500 g	= 1 lb 2 oz

Liquid Measures

30 ml	= 1 fl oz
125 ml	= 4 fl oz
250 ml	= 9 fl oz

Linear Measures

6 mm	= ¼ inch
1 cm	= ½ inch
2.5 cm	= 1 inch

Cup Conversions

1 cup plain (all-purpose) flour	= 125 g (4½ oz)
1 cup self-raising flour	= 125 g (4½ oz)
1 cup grated Parmesan cheese	= 100 g (3½ oz)
1 cup grated Cheddar cheese	= 125 g (4½ oz)
1 cup packaged breadcrumbs	= 100 g (3½ oz)
1 cup short-grain rice	= 220 g (8 oz)
1 cup mashed potato	= 230 g (8 oz)
1 cup sour cream	= 250 g (9 oz)
1 cup peas	= 155 g (5½ oz)

Oven Temperatures

Cooking times may vary slightly depending on the type of oven you are using. Before you preheat the oven, we suggest that you refer to the manufacturer's instructions to ensure proper temperature control.

	°C	°F	Gas Mark
Very slow	120	250	½
Slow	150	300	2
Warm	170	325	3
Moderate	180	350	4
Mod. hot	190	375	5
Mod. hot	200	400	6
Hot	220	425	7
Very hot	230	450	8

NOTE: For fan-forced ovens check your appliance manual, but as a general rule, set oven temperature to 20°C lower than the temperature indicated in the recipe.

International Glossary

beetroot	beets
capsicum	red or green pepper
chilli	chili pepper, chile
golden syrup	use dark corn syrup
red onion	Spanish onion
silverbeet	Swiss chard
soft brown sugar	light brown sugar

This edition published in 2003 by Bay Books, an imprint of Murdoch Magazines Pty Limited, GPO Box 1203, Sydney NSW 2001, Australia.

Editorial Director: Diana Hill. **Editor:** Rosalie Higson.
Food Director: Jane Lawson. **Food Editors:** Kerrie Ray, Tracy Rutherford.
Designer: Marylouise Brammer. **Recipe Development:** Tracy Rutherford, Denise Munro, Voula Mantzouridis, Alex Grant-Mitchell, Rebecca Clancy.
Home Economists: Melanie McDermott, Rebecca Clancy, Maria Gargas.
Photographers: Jon Bader, Reg Morrison (Steps).
Food Stylist: Carolyn Fienberg. **Food Stylist's Assistant:** Jo Forrest.
Chief Executive: Juliet Rogers. **Publisher:** Kay Scarlett.

ISBN 0 86411 367 6.
Reprinted 2004. Printed by Sing Cheong Printing Co. Ltd. PRINTED IN CHINA.